Praise for *Mapping Humanity*

"In *Mapping Humanity*, Joshua takes readers on a journey through the fascinating field of genetics. This book weaves together key scientific background information, exciting real-world applications, and important ethical issues in an accessible and engaging way."

—*Marie McNeely, PhD, cofounder of People Behind the Science and Unfold Productions*

"This book will serve as a very current and approachable way for the public to better understand this very interesting time in genetics and genomics, where the combination of sequencing and DNA manipulation technologies has brought us, while revealing the ethical and societal impacts."

—*Elaine Mardis, coexecutive director of the Institute for Genomic Medicine at Nationwide Children's Hospital*

Mapping Humanity

Mapping Humanity

How Modern Genetics Is Changing Criminal Justice, Personalized Medicine, and Our Identities

Joshua Z. Rappoport, PhD

BenBella Books, Inc.
Dallas, TX

BenBella Books, Inc.
10440 N. Central Expressway, Suite 800
Dallas, TX 75231
www.benbellabooks.com
Send feedback to feedback@benbellabooks.com

BenBella is a federally registered trademark.

Printed in the United States of America
10 9 8 7 6 5 4 3 2 1

Library of Congress Cataloging-in-Publication Data
Names: Rappoport, Joshua Z., 1974- author.
Title: Mapping humanity : how modern genetics is changing criminal justice,
   personalized medicine, and our identities / Joshua Z. Rappoport.
Description: Dallas, TX : BenBella Books, Inc., [2020] | Includes
   bibliographical references and index.
Identifiers: LCCN 2019059816 (print) | LCCN 2019059817 (ebook) | ISBN
   9781950665082 (paperback) | ISBN 9781950665259 (ebook)
Subjects: MESH: Human Genetics | Genetic Techniques
Classification: LCC QH431 (print) | LCC QH431 (ebook) | NLM QU 450 | DDC
   611/.01816—dc23
LC record available at https://lccn.loc.gov/2019059816
LC ebook record available at https://lccn.loc.gov/2019059817

Editing by Sheila Curry Oakes
Copyediting by Miki Alexandra Caputo
Proofreading by Lisa Story
   and Greg Teague
Printed by Lake Book Manufacturing
Indexing by WordCo Indexing Services, Inc.

Text design by Publishers' Design
   and Production Services, Inc.
Cover design by Pete Garceau
Cover image © Shutterstock / majcot
Text composition by PerfecType,
   Nashville, TN

Distributed to the trade by Two Rivers Distribution, an Ingram brand
www.tworiversdistribution.com

Special discounts for bulk sales are available.
Please contact bulkorders@benbellabooks.com.

This book is for Ben, Andrew, Madeline, Ema, and Kris.

CONTENTS

We are survival machines—robot vehicles blindly
programmed to preserve the selfish molecules known
as genes.

—Richard Dawkins, *The Selfish Gene*

Because the history of evolution is that life escapes
all barriers. Life breaks free. Life expands to new
territories. Painfully, perhaps even dangerously. But
life finds a way.

—Michael Crichton, *Jurassic Park*

The Power and Promise of Genomics

Everything that is alive is made of cells. Your body is made of many different types of cells—including tiny white blood cells that fight infection, thin nerve cells that conduct electrochemical signals, skin cells that protect internal organs, and muscle cells that produce movement. Although all the cells in your body serve different functions, every cell has one key thing in common: your genome.

Most of your cells contain a nucleus, an oval-shaped structure inside the cell, and the nucleus of those cells contain the bulk of your genome, a library of your DNA—everything that makes you you. While there are a few types of cells that do not contain a nucleus, and thus have little DNA (like red blood cells and platelets), these are still derived from precursor cells that have a nucleus that is lost during the final steps of differentiation.

DNA contains the blueprints that make up the main structural and functional components of our cells, our bodies, and ourselves. By studying the genome—a field known as genomics—we can hone our understanding of the fundamental mechanisms governing health, disease, development, and evolution. The similarities and differences among specific genomes help answer questions including how we evolved into the complex organisms we are and the genetic basis of diseases. Furthermore, genomic technologies are today being harnessed in applications ranging from the criminal justice system to food production.

For the last twenty years, I have studied cell biology. Cell biologists like myself devote their careers to understanding how cells work. There are different ways that cell biologists go about discovering how cells function:

- They mash up a bunch of cells and carefully quantify the various chemicals and molecules present—a bit like analyzing a milkshake to figure out how many strawberries went into it.
- They take cells and make them grow together into organized clusters resembling the organs found inside the body, creating so-called organoids that can be studied in culture; these organoids are like cells but demonstrate some of the more complex characteristics found inside the body.
- They use what we call "model organisms," such as fruit flies and mice, to study cells in the context of simpler and more easily manipulated living systems. This information can then provide critical insights into how similar processes unfold in our own bodies.

As a microscopist, I use high-powered microscopes to study what is going on inside cells. Generally, microscopists label specific components of the cell, and we often do this with living cells in order to study the dynamic functions of important parts of the cell as the processes of life unfold. But this can be a bit like learning to cook while only tasting the finished dish.

When it comes to food, you can gain a lot of information by examining the taste and appearance of the end product. It's obvious that mashed potatoes are better when they're smooth, rather than lumpy. You can taste the difference between bland, salty, and properly seasoned food. You can count the number of meatballs in your plate of spaghetti, and then break one open and see if it has onions inside. But to really understand how a dish is prepared, you need a recipe. The recipe tells you what goes into a particular dish, and how to prepare it. Without a recipe, how would you know how much garlic to add to the sauce, or when to add it?

To continue with our food analogy, you can learn a lot from viewing cells. But it's the genome that tells you about the building blocks of an organism and how it operates. This is why the genome is so important—it contains the recipe for the unique dish that is everything alive.

HOW THE GENOME WORKS

Basically, the genome inside our cells encodes the proteins that make up the structures of our bodies and perform many of the functions required for life. Individual genes contain a specific sequence of DNA that corresponds to a particular protein generated through a series of chemical reactions. The proteins encoded by genes are the main structural and functional components of our cells. Enzymes are proteins that perform specific functional roles—they can make a certain chemical reaction take place or act like a little molecular motor moving other proteins or structures around inside our cells. There are proteins that make our muscles contract, make our blood clot, make up our hair and fingernails, and perform countless other functions.

The genome provides the instructions for what the proteins should do, but it does not provide all the building blocks necessary for life. Some of the molecules needed for our survival—vitamins, minerals, sugars, fats, essential amino acids, water, and oxygen—must be brought in from the outside and incorporated into our diet. Other than the nutrients we ingest, nearly everything else that we are made of—our bones, hair, organs, and blood—is either formed directly from the proteins encoded by our genes, or as the product of enzymes encoded in our genomes. Even essential dietary nutrients must be stored, processed, and metabolized by proteins made via the instructions present in our genes.

Our genome contains discrete stretches of DNA called genes that contain the order of components that need to be put together to form specific proteins. It was once thought that each gene encodes exactly one, and only one, protein. However, it is now known that the information present in one gene can be processed many different ways, resulting in the formation of multiple distinct but related proteins. This would be like making cinnamon buns, hamburger buns, or a loaf of bread all with one type of dough that is treated differently depending upon the preferred final result.

All members of the same species will generally have the same genes, in the same basic order within the genome. However, humans carry two copies of each gene, one from the mother and the other from the father. Sperm and egg cells, referred to as gametes, have only one copy of each gene. During gamete formation, one from each pair is selected, and this is normally independent of any other gene. If one gamete has a person's maternal copy of gene 1, it generally holds that this has no bearing on whether the same gamete contains the maternal or paternal version of genes 2, 3, and so on. Thus, because the maternal and paternal copies of different genes mix and match to form the individual gamete, no two sperm or eggs are identical. This is like selecting which type of pasta, sauce, cheese, salad, or dressing that you pick from a buffet to put on your particular plate.

Later, when combined during fertilization, a matching pair for each gene, one from each gamete, is created. Although no sperm or egg contains exactly the same combination of genes from our mothers and fathers, on average these different versions of our genes are split equally into sperm and egg, and thus siblings typically share 50 percent of the same DNA. Humans have approximately twenty thousand genes, although estimates vary. Imagine that you and your sibling are standing at a very long buffet with twenty thousand different stations, each with only two choices, Caesar salad or house salad, chicken soup or minestrone, rolls or crackers, and so on. After the two of you went through the entire line of twenty thousand binary choices and looked at your (very large) plates, you should have selected about 50 percent of the same things, especially if the choices were random—and not influenced by dietary restriction or allergies.

Because our genes don't have the same DNA sequence, the corresponding proteins might be a bit different from each other because maternal and paternal versions of your genes are different. The exact DNA sequence makes my genome slightly different from yours. Although this difference can be subtle and reflect only a tiny proportion of the whole genome, this is what makes each of us unique.

Although every human will have the same genes, the variants of the DNA sequence in those genes are what creates diversity within the population. Ultimately, these variations primarily come from the inherent error rate in the enzymes that make copies of the genome when cells divide. These errors can have undetectable effects, cause subtle changes, or lead to catastrophic results, depending on the specific change in question.

The entire genome must be duplicated every time one cell splits into two cells, which is how we develop and grow on a cellular level. We start out as a single cell, a fertilized egg, which then divides countless times. Eventually, the cells differentiate and adopt specific characteristics that become the organs and cell types within the embryo and continue to develop throughout a person's life cycle.

CHANGES TO THE GENOME

The genome can change during cell division, and genomic variation that arises postfertilization can have effects upon the individual as well. This is the difference between an inherited mutation that was passed down from one of your parents and what is called a spontaneous mutation—for example, one that occurred early on in development when you were just that first single cell, or even as an adult. Any change present in sperm or egg will be passed down to the next generation. This is what it means for a mutation to be "heritable." However, changes that are not present in sperm or egg, such as a spontaneous mutation that arises during adulthood and causes a tumor to form at a particular place on our body, will not be heritable.

Changes to your DNA that increase the rate of division in a single cell can lead to cancer. If you get a really bad sunburn, the solar radiation can mutate DNA in your skin cells, potentially leading to skin cancer. These mutations, however, would not be passed down to your children. The reason you wear a lead apron when you have an x-ray is because the radiation used could cause mutations to occur, and if this happens in a sperm or egg those mutations could be heritable.

But not all mutations are bad. Heritable genetic variation leading to diversity in a population is the basic concept behind the theory of natural selection as proposed by Darwin. These variants can be called mutations, although that word has somewhat negative connotations. The inherent error rate of the enzymes that duplicate the genome during gamete formation or soon after fertilization ensures that a bit of change is introduced into the population. These variations can be positive, negative, or neutral for the individual, and the ultimate impact can depend upon specific environmental conditions, challenges, and stresses.

Random variation in the genome that is heritable and makes an individual better suited to a specific set of conditions will allow that person, and their progeny, if they too carry this mutation, to be fruitful and multiply, as well as to outcompete other individuals struggling against the same selective pressures. Thus, mutation is the currency of evolution. Although Darwin's theory of evolution—or more precisely, natural selection—is often referred to as "survival of the fittest," it can be more useful to think of it as selection of the mutations that provide the best chance for survival in a particular set of conditions.

One key aspect of the theory of natural selection is that it is not just about the success and survival of an individual who might carry a specific new genetic variant; the mutation in question must be heritable and make its way into the next generation so that it can potentially spread through the population. Therefore, in order for a mutation to be truly advantageous from an evolutionary perspective, it must ultimately promote survival of future generations, not only the individual. For example, if you can imagine a mutation that benefits the individual at the expense of the ability to procreate, this would not ultimately be beneficial to the species. In a similar fashion, mutations that improve the chances of mating, or the survival of progeny, would also be favored.

However, there is an additional facet of the theory of evolution that is referred to as sexual selection. This suggests that mutations that make an individual more attractive to potential mates will be selected, which accounts for physical features like peacock tails and behaviors like exotic mating rituals. These characteristics could be interpreted as indirectly demonstrating evolutionary fitness—for example, consider if only particularly healthy and well-fed potential mates may have the ability to display lush plumage or the energy to perform a mating display. But they could also have evolved owing to preferences in appearance or behavior in the potential mate. The cognitive basis for these types of preferences is an extremely active area of inquiry by evolutionary biologists. However, from the points of view of both natural selection and sexual selection, specific mutations that improve the ability of an individual to survive and reproduce will be selected and passed to future generations.

Mutations that cause disease or otherwise reduce fitness will similarly be selected against, if the effects result in a lower probability of successfully mating. However, mutations that cause diseases often don't emerge until well after mating occurs and therefore will not be subject to the same selective pressures. This is one way of looking at so-called diseases of aging and can help us understand why so many physical problems seem to emerge once we have passed through our reproductive prime. Of course, parenting and other benefits of maintaining older individuals in a family or group would provide evolutionary pressure to extend life span beyond reproductive age.

Ranging from cancer to heart disease and diabetes, the incidence of many diseases can be seen to rise sharply once we exit our reproductive years. One stark example of this phenomenon is Huntington's disease, a progressive neurodegenerative disease caused by a specific mutation that generally doesn't start to show symptoms until, on average, the age of forty. Thus, before genetic testing for Huntington's was available, it was hard to know if an individual might be affected and at risk of passing on the mutated gene during the reproductive years.

Generally speaking, many different and sometimes apparently contradictory forces can have effects on natural selection. Influences and circumstances can change, and a variation that might be advantageous at one time and place can be unwanted under different environmental conditions. The mechanisms that evolved to drive us to seek out salty, sweet, and fatty foods worked efficiently and appropriately when we were chasing down gazelle and searching for the occasional mango or handful of sweet berries devoid of bitter toxic chemicals. However, now that many people in the developed world live under conditions where rich and sugary food is easily attainable, and exercise rare, we are suffering as a species. Thus, the mutations that drove us to be better at converting these energy sources to fat were beneficial when these types of foods were scarce and our species was generally starved for sufficient nutrients, but now these same characteristics and traits have become potentially harmful given the differences in environment and behavior.

Of course, the genetic basis of the effects of the so-called Western diet—one that is high in sugar, salt, and fat—must also be considered. If two different people eat the same food, they won't necessarily process and store the nutrients in the same way. Over time, one person might develop diabetes, high blood pressure, or heart disease, but the other might not. From an evolutionary perspective, it stands to reason that people who are healthier, at least through reproductive age, through an inherent ability to better thrive on the Western diet, might be able to pass their genes on more effectively, and ultimately the human species as a whole will adapt. But even if this is the case, the process of evolutionary adaptation takes a very long time (many generations) and can result in the ultimately unnecessary suffering of many individuals.

Of course, not every species displays such long generation times. We can see evolution in real time when we consider antibiotics and bacteria. Penicillin is a chemical produced by bread mold, a fungus, and can kill certain types of bacteria. Over the course of their ancient molecular arms race with bacteria, certain fungi evolved the ability to produce antibiotics to gain an edge in the competition for nutrients. Penicillin and other antibiotics have proven invaluable in our own species' war against bacterial infection. However, through our overuse of antibiotics, we have altered the evolutionary history of bacteria.

If you expose a sample of billions of bacteria to antibiotics, there might be a few that are resistant. Because those bacteria harbor a genetic variation that makes them resistant to the antibiotic, they will survive while the majority will not. If the bacteria are maintained in culture conditions containing this antibiotic, ultimately only those with the antibiotic resistance will survive. However, it could be the case that the process of exhibiting antibiotic resistance takes a bit of extra energy or makes those bacteria in some way less robust. So if you were to remove the antibiotic and mix back in some of the original population, the antibiotic-resistant individuals would rapidly disappear, or at least decrease in number relative to those without the varation. This is because, as stated above, what is advantageous under one set of environmental conditions can be a disadvantage in another set of circumstances.

NATURAL SELECTION IN ACTION

Recently, the research group of Hopi Hoekstra at Harvard University published the results of an extremely exciting study that showed the entire process of evolution by natural selection in populations of wild animals. The group went to Nebraska and staked out three large areas with dark soil and three with light soil. They then cleared each enclosure of all the wild mice, collected groups of wild mice with a wide distribution of coat colors (from light to dark), took baseline DNA samples from the mice, and placed them into the enclosures.

Wild mice are prey for owls and other predatory birds that strike from above and look down to identify their prey. Contrasting colors—dark mice on light soil or light mice on dark soil—make the prey easier to spot, catch, and eat. Among wild mice, there is negative selection for coat color that contrasts with the

ground where a mouse lives and positive selection for camou-
flaged coat color that blends in.

The researchers then watched the mice over time to see what
happened. What they observed was extremely interesting and
represents a detailed demonstration of evolution in action. Eventually, they found that the mouse populations in the enclosures
changed color. As the darker mice in the area with light soil and
the light mice in the area with dark soil were more efficiently cap-
tured by predators, the populations shifted over time; the more
easily spotted mice were eaten, and the less obvious better cam-
ouflaged mice reproduced. Over several generations, there were
relatively more dark mice in the enclosure with dark soil and light
mice in the enclosure with light soil, as they were better able to
escape predation and mate. As it takes only about ten weeks for a
newborn mouse to become reproductively mature, these changes
can occur rapidly. Furthermore, the researchers demonstrated a
direct genetic basis for this observed change.

Coat color in mice is a complex result of several different
genes and proteins. The gene *Agouti* has been previously demonstrated to be involved in regulating mouse coat color, and variants
in the sequence of *Agouti* have been associated with differences in
the shade of the mouse's coat. At the start of the experiment the
researchers found that one particular mutation in *Agouti* that actually results in a functional change to the sequence of the agouti
protein was equally common among all mice in all enclosures.
This mutation alters the activity of the agouti protein in such a
way that the mouse's coat becomes lighter. This mutation then
became much more common in the light-colored mice found in
the enclosure with light soil, and less common in the dark-colored
mice in the enclosure with dark soil. Over time it spread through
the population in the enclosure with light soil, as the darker mice
were selectively predated, and similarly removed from the enclosure with light soil. Thus, the same mutation in the same starting
population was beneficial in one environment and detrimental in
the neighboring one.

In real time, these researchers took wild animals, subjected
them to different selective pressures in a natural environment,

and observed heritable genetic changes in the population over time that resulted in traits associated with better survival. Furthermore, the researchers were able to show that the two populations began to differ in a specific genetic variation dependent upon the isolation and differing selective pressures being applied, and that this genetic variation directly altered a trait associated with survival. The mutant agouti mice were lighter in color and selectively lost through predation in the enclosure with dark soil, while these same mice were protected from predation relative to the dark-colored mice in the enclosure with light soil. The populations that were not eaten were able to reproduce and have their genetics spread from one generation to the next.

Not all genetic variants lead to such clearly measurable differences in traits, but the more closely related to someone you are, the more of these variants you will have in common. When populations were more isolated, this meant that there would be more genetic similarity in one's immediate family than in one's extended family, and likewise more in one's clan or village than the broader region. If members of two nearby tribes never intermixed—like the mice in the two enclosures—over time they could develop very different genetic makeups, which might otherwise be surprising considering the relatively close physical distance of the two populations.

On a population level, this type of genetic isolation can have good consequences within natural environments. If a group of individuals has certain genetic variants in common that provide a specific advantage in the face of specific environmental challenges—such as the ability to effectively utilize a food source found in that area or to withstand a particular stress, such as extreme temperature or lack of water—they will thrive where others may not.

In wild populations under natural selective pressures, mixing across genetic backgrounds can cause these advantageous adaptations to be lost. If the two isolated populations of mice in Nebraska were allowed to mix, the differences in coat color might quickly disappear and, depending on the color of the soil, no longer confer a selective advantage against airborne predators.

Alternatively, some variations in the genome can have negative consequences, and when they accumulate in a population—for example, through inbreeding—without available alternatives, genetic disease can result. In nature, there will always be a balance between the benefits of genetic diversity reducing the risks of inbreeding, and maintaining specific adaptations in a population to respond to particular selective pressures.

GENETIC DISEASES

There are many different types of genetic diseases. Some are dominant, which means that one mutated copy of a gene will cause disease. Huntington's is an example of a dominant genetic disease. However, in many cases you won't display the symptoms of a genetic disease if only one copy of a gene is mutated. These types of diseases are referred to as recessive, and if you have a single mutated copy of a gene associated with a recessive disease, you are considered a carrier. Two mutated copies of the gene must be present to have the disease.

A mutant variant of a gene can persist in a population if it has no negative effect or if the odds of two carriers mating is rare. In an isolated population that only breeds within itself, the odds of inheriting two mutated copies from carriers breeding together increases. If the disease caused by combining two copies of a recessive mutation decreases the ability to successfully reproduce—for example, if it is debilitating or potentially deadly during childhood or early adulthood—that mutation might ultimately be selected out of the population. However, sometimes having a single mutated copy of a gene can actually be beneficial, and thus be maintained within a population.

Sickle cell anemia is a debilitating genetic disease that can arise when two carriers of a specific genetic variant mate. Those with a single copy of this variant display what is referred to as the sickle cell trait. They do not have the symptoms of sickle cell anemia, and there is actually a benefit to being a carrier. Having only one mutant copy of the gene in question, which encodes the protein hemoglobin that carries oxygen in our red blood cells, provides carriers with some protection against malaria, an infectious disease. For the unlucky people who inherit two mutant copies, this is a negative mutation. But for carriers it is a positive genetic variant. This mutant form of hemoglobin is found in some areas where malaria is common because the overall benefit of having the trait conferred by inheriting one mutant copy outweighs the relative risk of inheriting two copies and thus having sickle cell anemia.

THE IMPACT AND INFLUENCE OF THE GENOME

As humans have reshaped our environment, reducing natural threats and external stresses and pressures, and as interbreeding between individuals from populations previously insurmountably distant from one another has become a regular occurrence, genetic diversity among the human species has in some ways had the potential to accelerate rapidly. At the same time, however, when previously isolated members of a genetically diverse species mix, unique genetic characteristics can become blended together and disappear. Furthermore, the forces selecting for specific variants have shifted tremendously. Although evolution takes generations, the changes we have brought about to our daily lives have occurred incredibly rapidly on an evolutionary timescale.

Although the first human genome sequence was generated over fifteen years ago, it is still very early days for incorporating genomics into clinical practice. While gene therapy or personalized medical approaches are being developed and already having an impact on health and treating disease, we currently know much more about individual genetic variants than we do about how our genome as a whole is implicated in disease development. The ability to decipher the genome and understand the implications of particular variants within specific genes, as well as combinations of variants, including within different populations and environments, is rapidly accelerating. Genomics, the study of the genome, is fast becoming a critical tool in disease prevention, diagnosis, and treatment. However, the ability to apply the tools of genomics to individuals also raises significant questions of privacy, identity, and ethics. As with many technological advances, often concerns are only raised after the horse has already left the barn.

Our genes belong to us. Everything from our particular physical characteristics and medical information to our family ancestry to the tale of human history and the evolution of our species is written in our genes. They are the essence of who we are as individuals and a species. They are what makes us unique and what ties us together through common origins and biological constructions. Genes cannot be patented, as the Supreme Court ruled that each is essentially a "product of nature." That being said, researchers, medical professionals, and pharmaceutical and biotech companies are currently racing to develop gene therapies that can alter our genomes, hopefully improving our lives, reducing suffering, and prolonging life. Similarly, nearly every aspect of our food system is affected by genetic engineering. Scientists are even developing synthetic genomes that can be used to replace the natural occurring genomes in simple bacterial cells. Genomic technologies are fundamentally reshaping our understanding of identity and ancestry, and this information has significant implications for privacy and criminal justice.

The goal of this book is both to describe the power and promise of genomics and to accurately point out some areas for concern. The purpose is not to glorify and glamorize the genomic era that we are entering into as a panacea for all the ills of the world, nor to vilify the genomic researchers and corporations attempting to market and monetize genomic tools, techniques, data, information, and therapies.

Knowledge is power, and only through understanding the issues and specifics can informed choices be made. What is your genome? What makes it different from that of another human being? What are the implications of these differences? How can the genome be harnessed to help with medical issues, for both the individual and the population as a whole? What are the potential benefits and risks of living in a postgenomic world where anyone and everyone could have their genomes sequenced rapidly, cheaply, and accurately? This book will answer these questions, and will hopefully lead to insight into what it means to be human, what makes each of us unique, and what we all share in common.

I have written this book in a linear series of chapters that build sequentially. However, I have also aimed for each chapter to stand on its own so that readers can jump around or pick and choose chapters according to their particular interests. Although I have tried to avoid including long, redundant explanations in multiple chapters, critically important concepts are put into context whenever they are raised. Furthermore, the cross-references among chapters, as well as the index, should serve to provide any additional required explanations.

PART I

The Human Genome 101

CHAPTER 1

The Basics of DNA, Chromosomes, and Genes

The human genome is all your DNA, which stands for deoxyribonucleic acid. *Nucleic* in the name refers to the initial discovery of the vast majority of cellular DNA within the nucleus. The nucleus is an example of an *organelle*, a self-contained structure inside the cell that provides an isolated environment, contains particular molecules, and serves as the site for specific chemical reactions. However, a small amount of DNA is also present in your mitochondria, organelles that contain the machinery for generating the energy storage molecule ATP (adenosine triphosphate) from the metabolic by-products of digestion. ATP is the "universal currency" of chemical energy that powers nearly all aspects of cellular activity.

Mitochondria share a great deal in common structurally with bacteria, and it is believed that they emerged via a form of cellular symbiosis after bacteria entered an ancient cell (see p. 275). One characteristic that mitochondria share with bacteria is a circular chromosome containing several genes. However, this only represents a tiny fraction of the total DNA in our cells. The vast majority of our genome is present in nuclear DNA.

In the nucleus our DNA are forty-six chromosomes, long linear chains of DNA, half of which we inherit from each parent. Mitochondrial DNA, on the other hand, is a single large circle that is only passed down from the mother. While the sperm contributes nuclear chromosomal DNA during fertilization, our mitochondria only derive from the egg. Whether we are born male or female is determined by the combination of sex chromosomes we inherit. Females are XX, and each egg caries one X chromosome. Males are XY, each sperm providing either an X or a Y. All our other forty-four chromosomes are numbered, one to twenty-two, and exist in pairs. Excepting the two sex chromosomes, within each chromosome in a pair you find the same genes, generally in exactly the same order, although with minor differences between the maternal and paternal versions.

A gene is a stretch of DNA that will be expressed as a protein. Individual genes can be processed in different ways that can result in the expression of different versions of the protein. A protein is a chain of amino acids that has a particular structure and function in the cell or throughout the body. Variations among particular genes can cause changes to the corresponding protein structures they encode. These changes to the proteins caused by differing DNA sequence of the corresponding gene can have significant implications, particularly if an alteration in protein structure or function results. For example, there are genes that encode proteins that serve to promote cell division, the formation of two "daughter" cells from a single "parent" cell. This cell division occurs countless times during development, primarily *in utero*, but also as we grow up, and some cells continue to divide throughout adulthood. However, cell division is usually under very tight control. If variants of the proteins that drive cell division function in an out-of-control fashion, without appropriate regulation, it can lead to cancer.

DNA is a polymer, a long chain of regular repeating subunits, and the subunits that make up the DNA polymer are referred to as nucleotides. There are four different types of nucleotides in our DNA. Each contains an identical five-carbon sugar called ribose that has had one oxygen removed (thus *deoxyribose*), a negatively charged phosphate group, and then one of four so-called nitrogenous bases. These nitrogenous bases are referred to as adenine (A), thymine (T), cytosine (C), and guanine (G), and these single-letter codes are used to denote each individual complete nucleotide. Each of our chromosomes is basically one long chain of these four subunits; so a small stretch of DNA could read, for example, ATCGACTGTAGC.

However, DNA is generally found in the form of a double-stranded polymer. Stuck together, the two strands look sort of like a ladder, and the nucleotides present in one strand bind together with a cognate in the other: A binds to T, and C to G. The parts of the nucleotides that actually stick together are the nitrogenous bases, thus the pairs of nucleotides that bind each other across the two adjacent strands are referred to as "base pairs." The cognate base pairs for the sequence from above (ATCGACTGTAGC) would be TAGCTGACATCG. This two-stranded ladder is not straight; it twists around its central axis into a sort of spiral shape, like a piece of fusilli or a spiral staircase, and this is what is referred to as the "double helix."

The standard double-stranded reference human genome that includes every gene and all the other DNA found between genes contains about three billion base pairs (A-T, C-G). However, these three billion base pairs actually only include twenty-three chromosomes, one from each pair, and similarly only one copy of each gene, rather than the maternal and paternal pairs actually found in the nucleus of a cell. This is referred to as the "haploid" genome, or the number of chromosomes that would be present in sperm or egg. When chromosomes are paired up, this is known as our complete "diploid genome" (including all forty-six chromosomes and two copies of each gene), which contains about six billion base pairs of DNA. The human genome contains somewhere around twenty thousand different genes. However, a tremendous amount of our DNA is actually not in our genes but rather in regulatory sequences that can serve as binding sites for specific proteins that control gene expression, stretches of nucleotide sequences between genes that have no currently known function, ancient nonfunctional vestiges of genes that are no longer expressed, and many other forms.

Our genes are made up of long stretches of nucleotides that get converted into specific proteins by a series of enzymes and accessory proteins that control gene expression. The nucleotides are grouped together in triplets, called "codons," and specific combinations of three nucleotides are read out by the machinery that controls gene expression as different amino acids, which are the subunits of proteins. Analogous to the way that DNA is a polymer made up of a long series of nucleotides, proteins are a linear chain of amino acids. This specific conversion of a codon into one particular amino acid is what is referred to as the "genetic code." For example, the DNA sequence TGG tells the enzymes involved in gene expression that the protein to be created needs the amino acid tryptophan, which contrary to popular belief doesn't really cause us to feel sleepy after a big turkey dinner on Thanksgiving. Additionally, the codon GAA is for glutamate, which is the basis for MSG, monosodium glutamate.

Remember, DNA is actually made up of two adjacent stretches of nucleotides base-paired together, and the two strands in the double helix are different. One is referred to as the "coding" strand, which includes the information, the codons, that will be converted to protein, while the other is called the "template" strand, which is complementary to the coding strand.

Changing one nucleotide in a single codon can completely alter the amino acid encoded. Furthermore, there are specific "start" and "stop" codons that begin and end a protein sequence, respectively. A mutation that creates a stop codon in an inappropriate spot will shorten the overall length of a protein, and this can have disastrous functional effects. A mutation that introduces a stop codon that prematurely shortens the dystrophin protein is the cause of some cases of Duchenne muscular dystrophy.

Not all mutations arise from inappropriate stop codons. Even swapping one amino acid for another can have a significant impact on the structure and function of a protein. If that amino was particularly important—for example, in binding another protein or controlling a chemical reaction—one single change in sequence can cause disease.

Another type of mutation that can be particularly disastrous is called a "frameshift." This occurs when a single nucleotide in a gene is added or removed, which consequently alters every subsequent codon because they are read out in groups of three with no actual demarcation in the linear sequence. If you had a stretch of nucleotides that read ATCGTCATT, this is actually a group of three codons, ATC GTC ATT. If you removed the G, the sequence would now read ATCTCATT, and the second codon read would shift from GTC to TCA, swapping the amino acid valine to serine. Furthermore, if this occurred within a gene, every codon following that one would be similarly realigned as a result of the shift, and after that one point the whole protein sequence would change.

Genes are not directly converted to proteins, however; an intermediary is needed. Genes are present in the nucleus, within specific stretches of chromosomal DNA. They are then converted into ribonucleic acid (RNA) by a process known as transcription. DNA and RNA are quite similar in molecular structure. One feature that distinguishes DNA from RNA is that DNA (while chromosomal) is double stranded, and the RNA formed via transcription is single stranded. Another slight difference between DNA and RNA is that wherever the nucleotide thymine (T) is found in DNA, this is replaced by the structurally similar nucleotide uracil (U) in RNA.

Known as messenger RNA (mRNA), the sequence of the gene that is to be converted into protein is essentially a duplicate copy of the coding strand of DNA. The enzyme that makes an RNA polymer based upon a specific DNA sequence is referred to as RNA polymerase. It functions by sliding along the DNA template strand, which is temporarily separated from the cognate coding strand, and generates an RNA molecule that is exactly complementary to the template strand. This forms an RNA molecule that is identical in sequence to the coding strand, except that everywhere there was a T in the coding strand there is a U in the RNA sequence.

Once a final mature mRNA is formed, it exits the nucleus and then is fed into the ribosome, which is present in the cytoplasm outside the nucleus and is the site of protein synthesis. The codons are read in the ribosome and matched with the appropriate amino acid needed for the forming protein. This fundamental process of DNA being transcribed to RNA, which is then translated to protein, is referred to as the "central dogma of molecular biology" and elegantly encapsulates the key facets of gene expression. However, the story does not end there.

GENE EXPRESSION

The conventional understanding of molecular biology and gene expression is that chromosomal DNA contains genes, and these are transcribed by RNA polymerase activity into mRNA, which then exits the nucleus and is translated to protein by ribosomes in the cytoplasm. However, the classical concept of one gene, one protein woefully underrepresents the potential diversity of the human genome.

Although all our cells contain the same genome, tremendous diversity is observed. Neurons are remarkably different from muscle cells, and skin cells that protect us from external dangers are nothing like white blood cells that attack foreign invaders. One main source of this stunning variety is that not all genes are expressed at the same levels all the time. Some genes are expressed only in certain cells or tissues, or only following some specific stimulus, such as when a particular hormone is released. The onset of puberty, for example, is accompanied by huge changes in gene expression. Thus, regulation at the level of gene expression is responsible for a great deal of biological variability and change that can take place during development or when a specific cell enters a new phase of activity. However, beyond simple regulation of gene expression, there is a mechanism that greatly increases the total number of potential proteins expressed in human cells well beyond the approximately twenty thousand genes that have been identified.

A single gene can actually encode multiple different related proteins, and this ability is critical to increasing the diversity of characteristics displayed by our different cell types over time.

Some of the different versions of a protein encoded by a particular gene can be expressed in some cell types, whereas others can be present only under specific conditions. This is because the raw transcript that initially emerges from the RNA polymerase is not the final mRNA. One key process must occur before the final mature mRNA can leave the nucleus and be translated to protein: splicing.

A gene is not simply a linear sequence of DNA encoding a single protein from start to stop. Rather, genes are modular and are made of blocks of two different types of sequences, termed exons and introns. *Exon* is short for expressed sequence, and *intron* for intervening or intragenic sequence. Genes generally include many exons and introns, and following transcription the raw RNA is processed by splicing, which removes the introns so that only the exons remain in the final mRNA. A single raw RNA sequence can be processed multiple ways through what is referred to as alternative splicing.

When this splicing process occurs, different combinations of exons can be assembled from those present in the gene and raw RNA sequence. Imagine a gene with three exons (1, 2, and 3) and two introns (A and B). The gene and RNA would look like this: Exon1-IntronA-Exon2-IntronB-Exon3. Normally, for the full-length protein to be produced, the two introns would be removed by splicing that generates the final mature mRNA: Exon1-Exon2-Exon3. However, alternative splicing could remove the entire center of the RNA, including both the introns and Exon2, resulting in an mRNA that looks like this: Exon1-Exon3, encoding a truncated variant of the protein. As most genes contain many exons and introns, the potential for alternative splicing is immense. The triggers and regulatory pathways that control alternative splicing are very complicated but seem to be important for many aspects of development and a vast range of biological responses.

Sometimes cells are a liability and need to be removed. Like a double agent who was once trusted and now threatens a spy agency, an infected cell, or one that seems to be turning cancerous, must be eliminated efficiently and specifically. This is mediated by the process known as apoptosis, or programmed cell death. One way of inducing apoptosis is via the FAS pathway. All cells express a protein called FAS, which acts as a receptor that is triggered by another protein called the FAS ligand that is expressed by certain immune cells, such as killer T cells. When the FAS ligand on a killer T cell binds FAS on the surface of a target cell, apoptosis is triggered, and the potentially dangerous cell dies.

There are actually two different versions of the FAS receptor protein, generated by alternative splicing. The longer version includes the sequence that anchors the FAS receptor at the cell surface, making it available to be bound by the FAS ligand present on killer T cells. The shorter alternative version does not make it to the cell surface and is found within the cell, where it actually seems to be able to prevent apoptosis. Although the exact mechanisms regulating the alternative splicing of FAS are only now being uncovered, for the cell at least, which splice variant is expressed represents a truly life-or-death decision.

Beyond regulation of which specific genes are expressed at what levels, and which splice variants are translated to protein, there is a fundamentally critical process that is responsible for diversity among the members of a species, the actual sequence of DNA found in chromosomes. Changes at the level of DNA will be reflected whenever an altered gene is expressed, can affect wide-ranging regulatory processes if found outside gene regions, can be carried over following cell division and into the next generation following reproduction, and are responsible for complex outcomes, including adaptation through evolution as well as genetic disease. One single change among the billions of nucleotides in the human genome can have global consequences.

CHAPTER 2

Gene Mutations and Polymorphisms

If there was an error introduced during transcription, or if the ribosome makes a mistake, then an incorrect amino acid can be selected, which changes the sequence of a protein. This should be limited to that one particular cell at that point in time. However, a change in chromosomal DNA sequence will be reflected in all mRNA (messenger RNA) generated by transcription of that variant of the gene. Furthermore, this change can be maintained following cell division, and possibly even passed down to the next generation via reproduction. Not all changes to DNA are problematic. Some are totally innocuous, and others can, in fact, be beneficial. Mutation truly is the currency of evolution. When heritable changes to genes occur that alter the sequence of a protein in a way that confers some sort of advantage, this change will be passed on and ultimately spread through the population.

No two people have exactly the same genome. Of course, identical twins will essentially start out the same, but subtle changes throughout development can mean that, at least in some cells, differences between them will eventually occur. In general, different cells in your body will have slight differences in the exact sequence of some genes. This is how cancer can originate from a single aberrant mutant cell that outgrows its neighbors.

Each time a cell divides, the genome must first be duplicated. The enzymes that control cell division have a small inherent error rate. However, proofreading and DNA repair mechanisms exist to serve as quality control in an attempt to correct any mistakes. One of these fail-safes is that the two separate DNA strands are duplicated independently, so this means that an erroneously introduced nucleotide will not properly bind to the cognate complementary nucleotide in the adjacent strand. The proofreading machinery can look for these mismatches and then attempt to repair them. Tomas Lindahl of the Francis Crick Institute and Clare Hall Laboratory in England, Paul Modrich of Duke University School of Medicine, and Aziz Sancar of the University of North Carolina School of Medicine were awarded the Nobel Prize in Chemistry in 2015 for figuring out how DNA repair works. However, error correction in DNA sequence is not foolproof, and some mutations can make their way through these quality control mechanisms.

DNA MUTATION AND DISEASE

There are other ways that mutations can occur. Certain chemicals called carcinogens and types of radiation can alter our DNA sequence. Ultraviolet radiation from the sun's rays can damage DNA, and this can lead to skin cancer. Similarly, carcinogens in cigarette smoke can cause lung cancer. These environmental mutagens cause DNA damage that leads to these types of cancers, but this type of DNA damage is usually not heritable if it doesn't occur within sperm or egg.

However, many diseases are caused by mutations already present at birth that are inherited from the parents. In many cases, these genetic diseases occur only when mutated variants of a gene are inherited from both the mother and the father in what is called a recessive genetic disease. Broadly speaking, recessive mutations that can cause genetic diseases are rare in the general population and only become an issue if two individuals who are carriers for the mutation have a child together. This is more likely among inbred groups who are reproductively isolated from other populations.

If only one mutated copy is needed for the disease to occur, this is referred to as a dominant mutation. Generally speaking, recessive mutations decrease the functionality of a protein, so with one normal copy, enough residual function remains from the protein encoded by the nonmutated copy of the gene, and the disease arises only if two mutant copies are inherited.

Dominant mutations, however, often generate aberrant protein function that the unaffected copy of the gene cannot compensate for. A recessive mutation is a bit like when one side of your headphones stops working but you can still hear the music. Your headphones would only truly become nonfunctional if both speakers failed. A dominant mutation would be like one of the speakers starting to emit an annoying crackle or piercing noise, which would basically make the headphones nonfunctional. If a dominant mutation occurs spontaneously in an individual outside of sperm or egg—for example, in a skin cell—the effects of the dominant mutation can cause disease of a genetic origin during that person's life span, but such mutations will generally not be heritable. This is how many cancers—skin cancer, in this example—originate.

The epidermal growth factor receptor (EGFR) resides at the cell surface and binds to epidermal growth factor (EGF) that is present in the extracellular fluid. The binding of EGF activates EGFR, which sends signals into the cell initiating the process of cell division. Mutations in EGFR that lead to a protein that is inappropriately active—like a light switch that is stuck in the "on" position—and doesn't require EGF binding to stimulate cell division can lead to cancer as cells divide uncontrollably.

As we will see later, some cancers have a heritable genetic component. Breast cancer can run in families and is often associated with mutations in the BRCA genes. However, these effects are generally not apparent until later in life, and also often do not definitely cause the disease in all people that inherit them. These types of heritable mutations can display what is referred to as "incomplete penetrance," meaning that just inheriting the one mutant copy of a gene won't necessarily result in developing the disease. This isn't because the disease is recessive and two mutant copies are required. Rather, there are generally other genetic or environmental considerations. Maybe the gene in question is expressed only under certain circumstances, or other genes encode proteins that have functions that can regulate or compensate for the abnormal potentially disease-causing mutation.

There are also some heritable dominant mutations that will definitely cause significantly debilitating genetic diseases. These are said to display complete dominance. One example is Huntington's disease. As the disease is not evident until after the individual has passed reproductive age, this dominant mutation persists in the population. Because the disease doesn't affect reproduction, evolution won't select against this mutation, especially as innovations in sanitation and medical care have enabled humans now to live far longer than was previously possible.

So dominant mutations simply mean that one mutant copy is enough to cause disease. Dominant mutations can arise spontaneously and will generally not be passed down to the next generation if they do not occur in sperm or egg. Conversely, there is no natural selective pressure against heritable dominant mutations present in the genomes of all cells that don't display effects until after reproductive age. Furthermore, heritable dominant mutations with incomplete penetrance can cause disease in some people depending on other genetic variables, or behavioral considerations—for example, women with BRCA mutations can be at greater risk for cancer if they smoke, compared to women who smoke but don't have BRCA mutations, or those with BRCA mutations who don't smoke.

Another interesting type of genetic disease–causing mutation is carried on the X chromosome, referred to as X-linked, and may be recessive in females, meaning that with one mutated X chromosome and one normal, the disease won't be evident. In this case, even with a completely nonfunctional protein encoded by the mutant copy of the gene, the nonmutated copy on the other X chromosome will be able to compensate. Alternatively, the same mutation in a male, that only carries a single X chromosome, may result in the genetic disease. Some forms of muscular dystrophy, hemophilia, and even color blindness are X-linked, and are therefore much more common in males than females.

There are also some mutations in genes on the X chromosome that are so severe that even females display disease symptoms if they have one mutant and a nonmutant copy of the gene; this is known as dominant X-linked inheritance. In some cases, boys born with only the single mutant gene on their X chromosome won't survive. As will be discussed below, issues with X-linked diseases can be further complicated in females by the fact that one X chromosome is generally inactivated early in development, to prevent double doses of each gene being expressed in females compared to males.

All these types of so-called monogenic genetic diseases that can be caused by mutation in a single gene are less common than the ills that generally plague humanity, such as heart disease and diabetes. Most of these more prevalent diseases have a genetic component but don't stem from single mutations in individual genes. Rather, these are complex genetic diseases that are caused by the interplay of multiple subtle genetic variants and occur in the context of specific behaviors.

Some people are more prone to developing illnesses like heart disease or diabetes, and although there is a genetic component, depending on behavior and environment, the disease might not definitely arise. Similarly, specific conditions could occur where one person may develop a genetic disease but another person with a different genetic background will not.

POLYMORPHISMS

Although many fundamental concepts and methods in clinical and population genetics and genomics have been developed through the search for individual genes that cause a specific disease when mutated, more attention is currently being paid to large-scale studies of people who do and don't develop the more complex diseases with genetic components. In many cases, these studies are looking for small changes to many locations in the genome, none of which serve as definitive individual causes but together contribute to disease. Rather than mutations, these less well-understood variants are referred to as polymorphisms. In some cases, identifying combinations of polymorphisms associated with a particular disease can create a sort of signature for people who are prone to developing a disease.

Scientists also look deeply into specific polymorphisms, especially if they occur within a gene that could be involved in the development of a specific disease, or in a region nearby—for example, a location on the same chromosome as the gene of interest that could be involved in the regulation of expression of that gene. Proteins known as transcription factors bind to genomic DNA to regulate gene expression, and polymorphisms outside of genes that correspond to the binding sites for transcription factors can alter gene expression.

In addition to potentially providing information about protein function and gene expression, analysis of polymorphisms can be used for other purposes. Evaluating specific polymorphisms an individual carries is an effective identification tool, as no two people will have the same genome. This can form a type of genetic fingerprint with valuable applications in law enforcement, or to determine parentage, ancestry, and heritage.

DNA ANALYSIS

There are millions of potential single nucleotide polymorphisms (SNPs—pronounced "snips") in the human genome. These represent locations identified in the human genome where differences among people are found. SNPs don't necessarily occur within genes, and in many cases there are no measurable functional effects of particular SNPs. Genomic analyses in the context of ancestry research, for example, don't focus on individual SNPs but on the overall patterns among thousands of different polymorphic locations in the genome analyzed using advanced computational procedures. Generally speaking, the more closely related two people are, the more similar their specific patterns of SNPs will be. This extends beyond immediate family, and genetically isolated populations will also display more similar patterns of SNPs than people from different genetic backgrounds.

As geneticists collect more data and analyze the SNPs of more people from similar and distinct genetic backgrounds, an individual's genome can be used to more clearly understand that person's family heritage.

The Human Genome Project revealed the first complete reference sequence around fifteen years ago. Since then, the ability to completely sequence an entire genome has dramatically increased in efficiency and decreased in cost. However, as with the analysis of particular patterns of SNPs, you don't always need to know the entire sequence of the full genome. Sometimes you might only want the expressed sequences that get converted into protein, not including all the DNA in the nucleus that isn't found within genes—the exome, the combination of all exons present in the genome. If you are looking for mutations that change the function of a protein, you might only care about the sequences of genes that are expressed. Alternatively, if you already know something about a disease—for example, that it is caused by changes to a specific cellular pathway or process—you might specifically sequence only a subset of genes known to be involved in that aspect of cellular function.

So DNA sequencing does not only refer to one specific thing. Going back decades, biochemists and molecular biologists have worked out experimental procedures to define the sequence of individual short stretches of DNA. Over time, the length of DNA that could be sequenced has increased, and the speed and accuracy of the process has improved. However, this analysis has still generally been limited to single stretches of relatively pure DNA. More recently, the advent of what is broadly referred to as next-generation sequencing (NGS) has revolutionized DNA sequencing and made it possible to generate whole genome or exome sequences. Methods are first employed to enrich for the portions of the genome that encode proteins, the exons. Before exome sequencing can proceed, the exons have to be isolated and replicated relative to the rest of the genomic DNA.

Tools and techniques for DNA sequencing (discussed in the next chapters) can differ vastly in the maximal length of a single sequence that can be generated, in addition to accuracy, speed, and cost. Broadly speaking, generating a whole genome sequence is many times more expensive than exome sequencing. Of course, the more complex the data set, the larger the computer files and the more difficult to store and analyze. In fact, given how rapidly the costs of sequencing itself have been going down, it might ultimately make more sense to resequence a person's genome or exome as new biomedical questions and analytical strategies arise, rather than store the data for long periods.

CYTOGENETICS

There are also techniques that do not generate linear DNA sequence data but are employed to provide other types of genomic information. Cytogenetics refers to the study of how chromosome structure and function affect human health and disease, and for many years we have been able to identify each chromosome under a high-powered microscope and look for issues such as fragmentations or duplications. As demonstrated by conditions such as Down syndrome, which is caused by the presence of a third copy of chromosome 21 (also known as trisomy 21), these issues can be complex and debilitating. Thus, cytogenetic screening has generally been performed on early embryos, particularly in cases of advanced maternal age.

One main focus of genomic analysis is determining SNP patterns in individuals relative to a reference population. This analysis can range from the determination of SNP patterns associated with increased disease risk to identification of a particular individual, such as in paternity testing, to insight into ancestral heritage. These types of analytical approaches can be large scale across the whole genome or specifically targeted to particular genes, or even an individual SNP. A complete procedure ranging from sample preparation to data analysis also often requires information such as the ancestry or health status of the individual to provide a complete understanding of the particular question being addressed. Furthermore, each of these tools might require different analytical and computational frameworks so that significant results can be determined.

Taken together, our genomes hold a wealth of information that is rapidly changing our lives in diverse ways. Although having the human genome as a reference provides a great deal of power and flexibility, we don't always need all the information present in an individual's complete genome sequence if we can ask more targeted questions. You don't need to read the whole dictionary to find the meaning of a single word; you just have to know where to look. In the same way you can "search and find" a specific word or phrase in an electronic document or database, computational approaches and statistical tools are allowing investigations to be performed on wide arrays of genomic data sets, from deep full sequences of individuals to statistical analyses of thousands of people.

Ranging from research to medicine to law enforcement to ancestry, the applications of DNA sequence data are widespread. Hospital systems, universities, government agencies, and private companies are all obtaining and maintaining this data. This widespread use raises significant questions regarding privacy because, although your genome is unique to you, most of the variants that distinguish us as individuals are shared among close family members. This means that these types of genetic analyses have potential implications for the privacy of our relatives, something that is currently being actively exploited by law enforcement.

This is all happening now. You can have your genome sequenced for less than it costs to buy a used car, or even a nice bicycle. However, our understanding of these twenty thousand or so genes, not to mention any mutations or polymorphisms found within genes or other stretches of DNA, is greatly lagging behind our analytical power. Imagine that you could listen to someone speaking in another language and write down everything that person said—that's still a far cry from understanding the meaning of each and every word. So while we have a good understanding of many rare genetic diseases caused by mutations to specific genes that have profound functional effects, and are building databases full of SNP patterns found with a particular genetic background or associated with the propensity to develop a complex disease with a genetic component, the true power of genomic information is only slowly emerging.

ETHICAL QUESTIONS

With all this genetic information being discovered and stored, significant concerns regarding ethics and privacy must be considered. Since the power to identify an individual and to know medically relevant information, as well as family and cultural heritage, is all present in our genomes, with the advent of tools to permanently change genes—and to do so in a way that could be passed down to future generations—concerns that until recently seemed like paranoia or science fiction have suddenly become very real. That power is not limited to our own species. We can now interfere with the genomes of many other organisms, including the food we eat.

This is not necessarily cause for alarm, but it is certainly of some concern and at least should prompt us to inquire into the specific risks involved. Genomics is rapidly spreading through personal DNA testing, law enforcement, the pharmaceutical industry, agriculture, medicine, and many other areas. As consumers, as patients, as families, and as citizens, it is imperative that we gain an understanding of these important issues. In the pages below you will learn about gene therapy, genetically modified organisms, genetic genealogy, pharmacogenomics, genome-wide association studies, gene editing, and many other topics. This is the "brave new world" we are living in now, and we must understand the potential power of the genome, for good and ill.

Things have changed very rapidly in this area, and show no signs of slowing down. The first complete human genome took over ten years to produce, and cost billions of dollars. It is not actually the sequence from one individual but a composite of data from a number of volunteers. Moreover, it included just one copy of each chromosome, rather than two sets, and thus was a haploid reference, rather than a complete sequence. Although many applications (e.g., identifying criminals, understanding ancestry, and assisting in medical research and clinical practice) don't require a whole genome sequence, or significant sequencing of individual genes or large stretches of DNA, whole genome sequencing, exome sequencing, and studies of particular genes or chromosomal regions still require significant sequencing of chromosomal DNA. Thus, it makes sense to take a moment to understand how DNA sequencing is performed and how this has changed over time, and what different types of variables and concerns arise when considering DNA sequencing options.

PART II

DNA Sequencing Methods

CHAPTER 3

Sequencing DNA

As mentioned previously, DNA is a double-stranded polymer, and this structure is referred to as the double helix. It is made up of two separate side-by-side strands that bind to each other and then twirl around the shared central axis, like a spiral ladder. Each strand is a single polymer made up of a long chain of complementary nucleotides—As with Ts and Cs with Gs. If you have an A on one strand, you have a T on the adjacent strand; and where you have a C on one strand, you have a G on the other. This is referred to as Watson-Crick base pairing, or complementarity, especially when referring to a whole stretch of paired nucleotides. Therefore, if you had one DNA strand that reads ATGCTA, the complementary strand would read TACGAT.

On a nucleotide-by-nucleotide basis, the forces that bind the adjacent nucleotides within a single strand, the rails of the ladder, are significantly stronger than the forces that connect the two separate strands, the rungs of the ladder. That being said, when you have two very long DNA strands, the combined forces binding all the complementary nucleotides add together to make the two strands difficult to separate, rendering the double helix extremely hard to unwind. In fact, there is a specific enzyme referred to as DNA helicase that has evolved specifically to separate the two adjacent strands from each other, like sawing through the rungs of a ladder and splitting apart the two separate rails. The DNA polymerase enzyme moves along a single strand during replication; thus, as most methods of DNA sequencing rely on replication, separation of the two adjacent strands is generally the first step.

POLYMERASE CHAIN REACTION (PCR)

Helicase isn't actually required to separate the two complementary strands in a double helix. In fact, all that is needed is to heat up the DNA. Heating it to approximately 95°C, nearly the boiling point of water, is generally the first step in the technique referred to as polymerase chain reaction (PCR), which has been absolutely critical to developing techniques for DNA sequencing. PCR is the process used to make copies of a particular region of DNA. A polymerase is an enzyme that forms polymers. In PCR the enzyme in question is referred to as DNA polymerase. DNA polymerase generates a complementary copy of a single DNA strand. To be more precise, DNA polymerase extends a partial single complementary DNA strand along the paired complete template strand, or parent strand, matching the sequence base by base starting at what is called a primer. The primer is a short piece of DNA, usually around twenty nucleotides long, that is complementary to the parent strand being copied and serves as the starting point for the function of DNA polymerase to extend the new "daughter" strand. If the template strand read ATGCTGAGCG, and the primer used had the sequence TAC, DNA polymerase would extend the growing strand starting at the primer, first adding a G, and then an A, and so on until the complete complementary sequence of TACGACTCGC has been generated.

In addition to the DNA polymerase, the primer, and the parent strand being copied, the process also requires free nucleotides (As, Ts, Cs, and Gs) and a suitable source of energy to power the DNA polymerase, generally the high-energy storage molecule ATP. Together, these components are known as the reaction mixture. Additionally, the correct physical conditions, including temperature, are required. In fact, temperature is the critical factor when performing PCR.

PCR was developed around thirty years ago at the California biotechnology company Cetus by the late Kary Mullis, who later received the Nobel Prize for his work. The way that PCR works is very simple. You put the reaction mixture described above in a small thin-walled tube that allows rapid and efficient heat transfer and heat it up to about 95°C to separate the two parent strands. You then cool the mixture to about 60°C, and this makes the primers stick to the appropriate complementary spots on the parent strand, which correspond to the location where you want to make more DNA—the specific portion you want to copy. Rather than the parent strands reconnecting, the primers preferentially bind because you have a huge excess of primers in the reaction mixture compared to the amount of parent DNA. The final step of PCR is to warm the mixture up to about 72°C, where the DNA polymerase works efficiently to extend the primer with free nucleotides from the mix, and make more DNA following on from the position of the primer complementary to the parent strand, and identical to the other strand that was separated during the 95°C step.

Repeating this series of steps multiple times will result in the production of many copies of the sequence between the two primers. This duplication is possible because the DNA polymer possesses a property known as polarity. This doesn't have anything to do with electricity; rather, it means that the polymer has directionality—two ends that are different from one another. Each nucleotide, and thus the entire DNA polymer, has what are referred to as 5′ (five prime) and 3′ (three prime) ends. These names come from the fact that the core of each nucleotide is a ribose sugar containing five carbon atoms. The 5′ end is next to the fifth carbon in the ribose, and the 3′ end is by the third.

In a DNA polymer the 5′ end of one nucleotide in a strand binds to the 3′ end of the next nucleotide in the series. The entire molecule, even if it is the length of a complete chromosome, will have a 5′ and a 3′ end. Of course, DNA is normally double stranded, and the two complementary strands are connected along the entire length of the strand through Watson-Crick base pairing. The two polymers in the double helix are arranged in the opposite direction so that one goes from 5' to 3' and the other from 3' to 5'. This is referred to as the antiparallel nature of the double helix. So if a strand reads 5'-ATGCTA-3', the complementary strand would read 3'-TACGAT-5'.

DNA polymerase works in the 5' to 3' direction. It is only able to add the 5' end of a free nucleotide to the 3' end of a growing polymer, or a primer that is being extended. If you have one primer at the end of a specific sequence from a double-stranded parent molecule, and another at the other end complementary to the alternative strand, PCR allows you to rapidly create many copies of the region between the two primer locations. DNA polymerase will extend from one primer toward the other, and vice versa, so both primers will be extended and after several "cycles" of PCR the sequence of interest between the two primers will be amplified by the reaction. Each cycle of PCR involves the separation (95°C), primer binding (60°C), and polymerization (72°C) steps. If you start off with only one parent double helix, after twenty cycles of PCR you would have over one million copies of the specific DNA sequence of interest between the two primers, as each cycle duplicated the number of targets, from one to two to four to eight to sixteen to thirty-two to sixty-four and so on.

DNA polymerization had been described long before Mullis developed PCR and had been used to amplify specific regions of particular DNA sequences. However, the process was extremely challenging and laborious. Making copies of a specific sequence essentially involved creating the reaction conditions from scratch over and over again. The genius of the PCR approach is that once you set it up it will continue, as long as the limiting factors of the reaction mixture, such as the quantity of free nucleotides, don't run out. This ongoing process occurs because the temperature cycling works to ensure the proper steps continue. However, PCR was not possible as a true chain reaction until one critical piece was added.

Enzymes are proteins, and most cannot survive heating to 95°C. However, by using a DNA polymerase that was isolated from so-called thermophilic bacteria that grow in the near-boiling water of hot springs, the PCR reaction could cycle from one step to the other without having to add any further factors. The DNA polymerase most commonly employed for PCR is Taq, named for the *Thermus aquaticus*, a thermophilic bacteria first found in a hot spring in Yellowstone National Park. The addition of Taq was key to making PCR a practical reality. However, thermophilic bacteria and other organisms grow in hot springs in many locations, and other heat-resistant DNA polymerases have been employed in PCR as well.

Mullis originally had to manually cycle tubes of PCR reaction mixtures in different-temperature water baths around the lab, heating and cooling them in a laborious series of steps. However, now there are PCR machines in every biomedical research lab that automatically cycle the reactions through the temperature cycle. Furthermore, rather than using single test tubes, these PCR machines can hold plates about the size of two decks of cards next to each other that have many tiny wells that can each hold a single separate PCR reaction. These plates can hold 96 wells, 384 wells, or even 1,536 wells. The plate size is the same, but as the number of wells increases, the volume of each well gets smaller.

Although PCR isn't required for DNA sequencing, it makes it much easier. That being said, the overall perspective of using polymerization to duplicate a region of interest is essential to traditional DNA sequencing, regardless of whether PCR is employed. Termed chain-termination sequencing, or the Sanger method— after inventor Fred Sanger—sequencing DNA through polymerization was first performed in the mid-1970s. As described below, the name of this method stems from using special nucleotides that halt polymerization, and thus terminate the formation of the DNA chain. Sanger received the Nobel Prize for chain-termination sequencing in 1980.

THE SANGER METHOD

Whether through PCR or the older method employed by Sanger, the key to chain-termination sequencing is that a very small number of nucleotides are included in the reaction mixture that will stop the polymerization reaction. This means that at each location where a chain-termination nucleotide will be added, the polymerization reaction will cease. As a very small number of chain-termination nucleotides are added to the reaction, along with an excess of normal nucleotides capable of extending the DNA polymer, the odds of inserting a chain-termination nucleotide at any specific site are very low, and entirely random. The result will be many different polymers formed of different sizes, each with a different single chain-termination nucleotide at the end of the sequence. Following the procedures that have been developed and optimized, some polymers at every possible length could be formed.

The original Sanger method involved four separate reactions run in parallel, each with a different chain-termination nucleotide. One reaction contains a small amount of chain-termination As, another Ts, another Cs, and the fourth reaction had a few chain-termination Gs—let's call these A*, T*, C*, and G*, respectively. In all four reactions there are a number of normal As, Ts, Cs, and Gs, and in the first reaction there are a few A*s, in the second a few T*s, in the third a few C*s, and in the fourth a few G*s.

These four polymerizations reactions were performed separately and then analyzed. If the goal were to amplify the sequence TAGCGAT, the reaction with chain-termination A*s, in addition to all the normal As, Ts, Cs, and Gs, would result in three different products, of which one would just be the complete sequence without any chain-termination nucleotides added, TAGCGAT. Then there would be only TA*, with no following nucleotides, if the A at the second position was an A* chain-termination nucleotide, and finally there would also be TAGCGA*. If this process were repeated with all four different reactions the result would be a mixture of chain-termination nucleotides, each at different possible positions in the sequence. The trick then is to figure out the positions for each; this tells you the sequence, which is determined by measuring the length of the different DNA polymers.

DNA length can be measured quite accurately simply by physically separating the different polymers. The standard way this is done is by injecting the DNA into a slab of agarose, a gel that comes from seaweed prepared in a salt solution that can conduct electricity. DNA is naturally negatively charged, so if you apply an electric field to the gel, the DNA will move away from the negative pole, where you inject the DNA, and migrate toward the positive pole. The smaller DNA moves faster than the larger pieces of DNA, so the DNA gets separated out by size. With a mixed population of DNA strands of specific defined lengths, there will be characteristic "bands" of the different groups of molecules, each corresponding to many molecules of polymers of identical length. If there were a number of DNA polymers that were fifty nucleotides long, another group that was one hundred nucleotides long, and finally some that were two hundred nucleotides long, there would be three discrete bands in the gel, each corresponding to one of the groups, and the shorter molecules would migrate farther down the gel toward the positive pole while the longer ones would stay closer to where they were inserted into the top of the gel, by the negative pole. If you separate out all the fragments of various sizes in a gel, with the four different chain-termination reactions side by side in parallel lanes in the gel, the end result will be four separate patterns of bands, each showing the length of a DNA strand that ended with one of the different chain-termination nucleotides. The four different reactions are injected into distinct positions at the top of the gel and then migrate from the negative pole to the positive, like swimmers in a race moving through their respective lanes.

The different tracks followed by the specific samples are in fact referred to as lanes. For the sequence TAGCGAT, from the bottom of the gel up, the first band would be in the T* lane (just the first T alone), then one in the A* lane (TA), and then one in the G* lane (TAG), and so on. As the single T* would run the fastest in the gel, this would show that the shortest polymer ended with a T, and thus that was the first nucleotide in the sequence. The next shortest possible polymer would be T plus an A*, and that would mean there would be a band migrating just a bit slower than the fastest in the A* lane, corresponding to TA*. The third-fastest molecule would be TAG*, and so on until the complete sequence is completed.

Performing four separate reactions and running each product on a separate lane of a gel and then interpreting the results is extremely inefficient and time consuming. By using PCR, the process was greatly accelerated. Furthermore, one extra development allowed for even higher efficiency sequencing by modifying the chain-termination nucleotides. Specifically, this involved using fluorescent dyes to label the different chain-termination nucleotides, using a different color for each. This is referred to as dye-termination sequencing, in which every strand that ended with an A* would be one color, every T* a different color, and so on. With this approach, the four chain-termination nucleotides could be combined into a single PCR reaction, and the results could be analyzed together. This meant four separate lanes of a gel weren't needed because four different pictures of the gel could be taken, each with different filters that would selectively image each separate color. More advanced methods for chain-termination sequencing with fluorescent nucleotides don't use gels; rather, the different polymers are separated by size in tiny capillary tubes and illuminated with different-color lasers, further increasing efficiency.

Sanger sequencing, or dye-termination sequencing with PCR, can only be used to sequence DNA polymers of less than about a thousand nucleotides long, and this is as long as you can generally make a PCR product. However, many longer DNA targets need to be sequenced. Longer sequences are analyzed by breaking up the longer polymers into smaller pieces, sequencing each, and then combining them together. This method is referred to as "shotgun sequencing" and served as the basis for the Human Genome Project. The key to this technique is that the original parent polymer is broken into multiple overlapping sections and then each one is sequenced separately. You can then construct what is referred to as a "contig," a single overall sequence that aligns all the smaller segments based upon smaller regions where overlap is found.

For example, if you had a tube full of a DNA polymer that read TAGCGAT, and that was broken up into smaller overlapping sequences such as TAGC, GCGA, and CGAT, you could then combine these based upon the regions of overlap to return to the full-length sequence of TAGCGAT. If the longest single sequence you can get from the Sanger method is about a thousand nucleotides, and the shortest human chromosome is about fifty million nucleotides in length, you can begin to appreciate the unbelievably huge number of individual sequencing reactions needed to generate the first complete human genome sequence.

Thankfully, more automated and higher-throughput techniques are being developed all the time. These are referred to as "next-generation sequencing" and serve as the basis for a vast number of innovative advanced genomics tools that would otherwise not be possible.

CHAPTER 4

Next-Generation Sequencing

A s stated in the previous chapter, the general idea behind Sanger DNA sequencing is that by using a small number of fluorescent nucleotides that arrest a PCR reaction, chain termination results at different positions among the newly synthesized molecules. This creates a mixed population of PCR products that together represent every possible length along the target DNA sequence of interest. Each piece concludes with the final fluorescent nucleotide carrying a different color corresponding to A, T, C, and G. By effectively measuring the order in which these fragments separate by size, and by knowing the specific fluorescent chain-termination nucleotide on the end of each, you can deduce the entire linear sequence of the complete sample being sequenced. Sanger-based sequencing can involve separating and analyzing huge numbers of PCR products. With larger targets, such as an entire human chromosome, even a full-length PCR product will correspond only to a tiny fragment of the complete DNA molecule you are trying to sequence. By using shotgun sequencing, the target is fragmented and then Sanger sequencing is repeated for many overlapping portions of the entire genome. These fragments then must be combined in the correct order to create the entire target sequence.

Most standard PCR reactions are limited to around a thousand nucleotides. Given that the complete diploid human genome includes six billion base pairs, performing whole genome sequencing (WGS) via shotgun Sanger sequencing is a truly astounding feat. In fact, it took hundreds of scientists over a decade, at a cost of around $3 billion, to complete the Human Genome Project in 2003, which was limited to a haploid reference genome that "only" included the three billion base pairs present within a single copy of each of our twenty-three chromosomes.

Since then, multiple new approaches to DNA sequencing have emerged. These vary in the specifics, but each tries to improve overall efficiency. Most employ some innovative biochemical trickery in an attempt to rapidly generate many short overlapping sequences from a large target of interest. Together these have been referred to as next-generation sequencing (NGS). However, for reasons that will become clear, collectively many of these methods are classified under the term massively parallel sequencing.

There are many ways to compare different sequencing techniques. You can focus on financial issues, such as how much it would cost per run, per sample, or per genome. Alternatively, you can estimate speed or efficiency through measures such as how many nucleotides per second can be sequenced. This is a key issue, especially with WGS, and raises the question of how, instead of taking over a decade, sequencing a whole genome currently takes days or even hours. How can it be possible that rather than measuring WGS costs in the billions of dollars, we now discuss techniques that can cost less than $1,000 per genome?

Finally, when comparing different DNA sequencing platforms, estimates of error rates must be considered. The question of error in DNA sequencing is actually quite complex, as it can arise at any point of the entire process. However, as will be discussed in more detail below, analysis of the potential sources, and how much error might be acceptable for a particular application, must be performed in the specific context of the questions being addressed.

WHY PERFORM DNA SEQUENCING?

There are many different types of applications for DNA sequencing, and there is no clear definitive winner among the different techniques available that is significantly better in all ways and ideal for the different types of analyses required. Identifying a virus from a clinical sample might require fast sequencing of a relatively small genome, and this might not need to be done with extremely high accuracy. Many viral genomes are only a few thousand nucleotides long and can be quite easy to identify. A single Sanger sequencing reaction might be suitable for this work. If you were performing epidemiological work in the field, such as studying an outbreak of Ebola in Africa, the key criteria for your ideal sequencing technology might be speed, ease of use, and cost, rather than focusing on error rate, or length of raw sequence produced. Alternatively, if you are investigating the genetic basis of a disease that arises in part from the combined effects of many subtle variants, searching for rare SNPs in human WGS requires very accurate analysis of long genomes. Furthermore, often NGS technique can be followed up with other confirmatory approaches, such as Sanger sequencing—for example, to verify novel SNPs identified with an initial more high-throughput method.

NGS TECHNOLOGIES

One thing many NGS technologies have in common is that rather than involving analysis of PCR products after they have been produced, as in Sanger sequencing, several of the methods for massively parallel sequencing actually deduce the sequence of the target of interest in real time as polymerization occurs, one nucleotide at a time. This is referred to as sequencing by synthesis, and although there are a few specific ways of determining which nucleotide is being added to the growing DNA polymer, in many ways these techniques are quite similar to one another.

Overall, going from a genome, which is a physical thing, to a sequence, which exists as a computer file, involves three general steps: sample preparation, sequencing, and data processing. Although simply logging A, T, C, or G may seem like a small task for modern computers, doing this millions, or even billions, of times, and including quality control information—basically, the certainty that the sequence you obtained is correct—really adds up. Furthermore, this is just for the raw sequence files, before any processing or analysis. Usually processing includes combining the raw sequences of different fragments together and aligning them to a known reference genome in the correct order. Data analysis, such as comparing two different genomes, or predicting the potential functional consequences of a specific variant, such as how differences in DNA sequence might alter protein structure or function, can be considered a separate set of procedures that can be performed once the complete sequence has been determined. This type of analysis generally involves making use of any number of extensively well-curated international molecular databases—for example, to find out if anyone else with the same variants might have been previously identified and studied medically. Ultimately, however, it is in these three phases of sample preparation, sequencing, and data processing that the primary significant differences among different sequencing technologies emerge, as do the errors that might affect data analysis.

In most types of massively parallel sequencing the first step is that the DNA being analyzed is broken up into small fragments. As in shotgun Sanger sequencing, long starting templates are broken up into smaller pieces, which are then sequenced individually. These fragments, and the individual sequencing "reads" in massively parallel sequencing, are generally much smaller than in Sanger sequencing, and thus there are also correspondingly many more of them. This provides a much higher-throughput approach overall but also increases the complexity of the data processing required.

There are different ways this initial fragmentation can occur, but usually it involves either enzymes that cut up DNA into small pieces, or physical means—sort of like putting the DNA into a food processor. Following fragmentation, generating many short PCR products can form a collection that includes multiple copies

of each small piece of the total target. The prepared DNA ready
to be sequenced is referred to as a library. Having many identical copies of each fragment in the library increases the signal
that will be detected as the sequencing reactions occur. As we
will see below, this signal is generally light emitted from a fluorescent nucleotide, as in Sanger sequencing, or detection of an
electrochemical current. As each small template PCR product to
be sequenced in the library might only be a few hundred nucleotides long, in order to sequence something like a whole human
genome, which is six billion base pairs, libraries including a truly
staggering number of individual fragments must be analyzed in
parallel. Additionally, enough distinct overall fragments must be
present in the library to permit overlap of each individual section
with others. All this information then must be combined to make
the final overall linear sequence of the whole target, employing
extremely sophisticated computational approaches—and, if possible, alignment to a standard reference sequence, generally a previously determined sequence from the same organism, or one
that is closely related.

Another aspect of this type of NGS is generally going to
involve some kind of physical separation of the different individual templates in the library. Many small overlapping PCR products generated from the sample to be sequenced can be physically
and spatially separated from each other. One way this can occur
is by binding each fragment to a tiny bead, literally a small plastic
bead that has the ability to stick to DNA. Conditions are employed
to ensure that only one DNA fragment binds to each bead. Each
bead is physically separated and can then serve as many individual locations for sequencing by synthesis, giving a very large pool
of unique separated fragments, each ready to serve as the template for a tiny portion of the whole target. Once there is a library
that corresponds to many physically separated fragments, these
are then replicated multiple times to increase the ultimate signal
when proceeding with whatever specific technique for sequencing has been selected.

In many ways, it is the specific method of detecting the identity of the nucleotide being added—whether it is an A, T, C, or

G—that distinguishes the different techniques for sequencing by synthesis. The most straightforward technique resembles Sanger sequencing but differs in one critical respect. This technique, called reversible dye terminator sequencing, also employs fluorescent nucleotides that arrest the PCR sequencing reaction, but then modifies these nucleotides in a controlled fashion to permit elongation of the DNA polymer to continue.

The general idea behind reversible dye terminator sequencing is that different steps are performed in a cycle to identify the sequence one nucleotide at a time in real time as they are added. Through the use of reversible dye terminator nucleotides, one nucleotide is added; then it's determined what it was; the chemical block preventing addition of the next nucleotide is removed; and then the next nucleotide is added. Ultimately these cycles are run until PCR reactions producing complementary copies of each individual template fragment are completed, and each nucleotide has been identified immediately after it was added.

As the templates are physically separated from each other to start with, a camera is used with filters that can separate the four different colors associated with the dye-labeled nucleotides. The camera takes pictures to record each nucleotide addition step, and this results in the generation of a huge folder of image files showing tiny dots that change color over time depending on which nucleotide is being added. Each dot corresponds to a different spatially distinct template fragment at a known and fixed location, and the color changes are dependent upon the nucleotides being added one at a time in each cycle, which in turn is based upon the sequence of the target, as nucleotides are added that are exactly complementary to the template.

The key technological developments behind reversible dye terminator sequencing include the physical separation of the different template fragments and the use of the reversible dye terminator nucleotides. However, this technique would not be possible without huge advances in microfluidics. Basically, this means that the sequencing reactions can take place in a physical environment where very small volumes of fluid can be added and removed extremely quickly and precisely. This allows for the ability to rapidly pulse the dye terminator nucleotides into the reaction mixture, which is followed by a single image that reads out which color is bound at each site. Then the free unincorporated nucleotides are removed and chemicals are added that alter the dye terminator nucleotides so that they are no longer fluorescent and can now accept the addition of the next nucleotide in the chain. New reversible dye terminator nucleotides are then added; one is incorporated; a composite image with all four colors is collected; and the cycle continues.

There are several other techniques for sequencing by synthesis. In pyrosequencing, a flash of light is generated each time a nucleotide is added to the growing DNA strand. Instead of being caused by a tag carried by the nucleotides, the flash of light occurs owing to a chemical reaction coupled to the addition of nucleotides to the growing DNA polymer as synthesis progresses. There is no discrimination of nucleotides according to specific dye color; rather, the microfluidic system is able to selectively pulse in each different nucleotide separately in a known order, and then wash away all the free, unincorporated nucleotides before the next is added. As with reversible dye terminator sequencing, a camera captures the images of the physically separated templates as one nucleotide after another is added, only in this case there is only one image collected per cycle with no color information because only the spots that incorporated that specific nucleotide will show a flash. The key is knowing which nucleotide has been added to each template by determining whether a flash of light is detected at that physical location after adding each of the four in the series. A few years after the Human Genome Project, pyrosequencing was employed to generate one of the first human genome sequences successfully completed through next-generation sequencing. This took only two years and cost approximately $1 million. The reduction in time and cost was a huge advance compared to the first genome compiled.

Quite similar to pyrosequencing is ion semiconductor sequencing. Rather than a burst of light being generated, a small electrochemical change is detected every time a nucleotide is successfully added. The templates are physically separated in such a way that these electrochemical changes can be spatially localized, and the different nucleotides are pulsed into the system in a known order. Although the ultimate readout might be different among these specific methods (a particular color of a fluorescent molecule, a burst of light, or an electrochemical change), the overall strategy is essentially the same.

One interesting issue with both pyrosequencing and ion semiconductor sequencing is that both can have difficulty reading out stretches of identical nucleotides. Because each single type of nucleotide is added to the sequencing reaction chamber one at a time, synthesis is paused by switching from one nucleotide pool to the next, rather than by reversible chain termination. Thus, if you have a long stretch of DNA that contains the same nucleotide in a series, it can be hard for the DNA sequencing instruments to distinguish exactly how many individual flashes of light or electrochemical changes took place during that cycle. Although it can be straightforward to determine if there were one or two As, Ts, Cs, or Gs added to a particular template in one cycle, figuring out if seven or eight of the same nucleotide were together in a stretch might not be possible.

Imagine if people were singing and you were blindfolded: Although it would be very easy to tell if a solo singer was joined by one other person and was now singing a duet, do you think you could tell by ear if there were seven or eight people singing together in a group? Generally speaking, as the difference between signals decreases, the ability to precisely determine the number becomes more difficult.

This is not the only issue with analyzing regions containing regularly repeating sequences of nucleotides found with massively parallel sequencing. It is not clear exactly why, but besides protein-coding genes, genomes have huge numbers of repeating regions, some long and some very short. These can include a few nucleotides in the same order, with that small motif repeated over and over, or can contain other types of repeated sequences. The number of nucleotides per individual sequencing reaction, known as the read length, of each specific isolated fragment in massively parallel sequencing is quite short, from tens of nucleotides to generally a few hundred at most. If you have the same sequence repeated over and over, it can be extremely difficult to ultimately place the sequences of these fragments into the correct order. Having complete reference sequences, such as from techniques generated with longer reads, like shotgun Sanger sequencing, greatly assists the alignment of repeated sequences. However, given that repeated sequences can vary from person to person in the number of times a specific sequence is repeated, many WGS analyses essentially fail to capture some genomic variability.

More recently, other sequencing techniques have been developed that provide extremely long reads. One such new approach is referred to as nanopore sequencing. Rather than huge numbers of short reads performed in parallel and then combined, this technology can generate sequences of extremely large targets at a stretch, even up to millions of nucleotides long. Many fewer individual reads are required to generate a complete genome, and with small genomes, such as viruses, the whole sequence can theoretically be completed in a single read. However, the error rate in generating the specific sequence is quite high.

The way nanopore sequencing works is that the DNA of interest is fed through a small channel present in a membrane within the DNA sequencing device at predetermined locations. The membrane is very thin and acts as a barrier to ion flow from one side to the other. When a voltage is applied across the membrane, these channels normally allow ions to flow through them, like little pipes through which saltwater can move, and this ion flow can be measured as tiny electrochemical currents at the sites of each channel. When the DNA polymer enters the channel, the DNA reduces the flow of ions being measured. This is like a pipe with a partial clog. Imagine, then, if you could tell what was in the pipe by measuring how much water is still able to flow. This is possible because the reduction in the flow of ions is not complete, and the extent of change in current can be measured. This will depend upon the exact nucleotide sequence present within the channel at that time point. Thus, by measuring the changes in current as the DNA polymer feeds through the channel, you can associate specific changes with particular nucleotides.

You might picture this process as Ping-Pong balls (ions) flowing through a hollow pipe and then adding a rope with tennis balls attached to it (DNA). The pipe is big enough to allow passage of both types of balls, but the flow of Ping-Pong balls would be slowed by the presence of the tennis balls. Now imagine that in addition to tennis balls you also have some other types of balls on the string. Baseballs are slightly larger than tennis balls, and softballs are much bigger. So if you add these on the string as well, you could look at the flow of Ping-Pong balls and figure out which types of balls were inside the tube at any one moment. The larger the ball on the string that is in the tube at that moment, the slower the flow of Ping-Pong balls that will be measured.

The devices that have been developed for performing nanopore sequencing are very small, modular, and simple to use. As the target template doesn't generally need to be fragmented, there really is no library creation step similar to what is required in massively parallel sequencing. The DNA polymer to be sequenced simply has to be isolated and introduced to the channels in a controlled way. This usually involves separating the double-stranded template into a single strand, as only single-stranded DNA can feed into the nanopore channel. Once this occurs the sequence readout can commence.

ERRORS IN SEQUENCING

Errors in DNA sequencing can exist in several forms, and can have different root causes. Every DNA polymerase has an inherent error rate. Occasionally the wrong nucleotide will be added, and it won't actually be complementary to the template. However, many copies of each small template fragment will be associated with each specific location in the sequencing data set. If only a few errors are introduced during sequencing by synthesis, as long as the majority are correctly duplicated, the appropriate nucleotide will be logged for that specific position.

However, what if the error was introduced during the preparation of a library? If that occurs, an incorrect nucleotide might be assigned corresponding to a position within that particular template fragment. This can actually be a very easy source of initial error to deal with. All that is needed is to sequence the same target multiple times using different specific template fragments. The library preparation step generally produces many different overlapping fragments of the target being sequenced, and because there are many ways to break up the template, each nucleotide in the overall sequence will be present in multiple independent reads. This oversampling is referred to as the extent of "coverage" in your DNA sequencing protocol; it is essentially how many times each individual nucleotide was independently sequenced through analysis of many distinct overlapping stretches.

As they say, practice makes perfect. Even if you are a very accomplished pianist, you will occasionally make mistakes when playing a very technical piece. However, if you were to play the same piece five, ten, or even fifty times, recording each attempt and then combining only the best-sounding notes in the final version, you would have a perfect representation of the song you were trying to play.

In next-generation DNA sequencing, especially massively parallel sequencing by synthesis, the huge number of individual fragments facilitates repeated sequencing to increase coverage. The exact number of times you need to sequence each nucleotide depends on the specific technique employed, as well as the ultimate questions being asked. The shorthand notation employed to quantify the extent of sequence coverage—how many times a sequence was generated for each individual site—is referred to with the notation of "×" so that 5× coverage means five times, and 10× ten times. Although, depending on the technique and application, specific recommendations vary regarding coverage, generally for WGS something like 30× coverage is considered sufficient for most purposes. But there are some other sources of error in DNA sequencing that are not solved simply by repetition.

Overall, the sources of error in DNA sequencing—apart from inherent issues with DNA polymerization—can be split into the three broad phases of the process: sample preparation, the sequencing reaction (template DNA, nucleotides, primers, and DNA polymerase), and data processing. There are other issues that cause problems in many techniques for massively parallel sequencing during the process of library generation beyond simple polymerization errors. Some regions of a genome might be easier to use as templates to generate PCR products during library preparation than others. This can depend on the specific sequence of the region, which can affect how easy it is for the DNA polymerase to do its job. Often, the library that is created doesn't exactly represent the entire target that is being sequenced, such as a complete human genome. In this instance, increasing coverage—basically, repeated sequencing—will not improve your results. If the library doesn't contain certain regions of the genome, these won't be present in the final sequence.

Beyond the general issues with incorrect incorporation of nucleotides by DNA polymerase during PCR, the actual sequencing reaction can be a source of other types of error. Another issue is what is referred to as incorrect "base calling." (Base calling is essentially the identification of nucleotides as the sequencing reaction progresses.) Errors occur when the interpretation of the signal being read out during sequencing is incorrect, like pulling a numbered Ping-Pong ball out of the hopper and reading out the wrong position in a game of Bingo. When that happens, the template is present in the library (and has the correct order of the right nucleotides), but for whatever reason the signal that is collected—whether it is a specific color in an image, a burst of light, or an ionic change—is misinterpreted and the wrong nucleotide logged. One possible cause of base-calling errors might be if the locations of the sequencing-by-synthesis reactions are too physically close to each other relative to the resolution of the system to be effectively discriminated.

Manufacturers of the different instruments employed to perform the above-described next-generation DNA sequencing techniques have recognized the importance of this physical spacing in being able to effectively separate out the different sites of synthesis. In most modern machines the sample chamber has been microengineered so that the locations of the sequencing reactions are tightly controlled and optimally spaced to incorporate as many parallel reactions as possible but maintain sufficient spacing to prevent physical overlap in signal readout.

Other issues in the sequencing reaction can cause problems as well. In reversible dye terminator sequencing, if the dye isn't removed properly from a template before adding the next nucleotide, two different signals might be detected at one spot. If the chemical block preventing addition of the next nucleotide doesn't function properly, there can be two nucleotides added at once. If removing the chemical block fails, that specific template won't get extended as the reaction progresses. Given that each physical spot contains many copies of the same template, these types of issues are not catastrophic, but over time they can affect the signals detected at a specific site and cause potential confusion in base calling.

Another significant challenge occurs during data processing and analysis, employing specific software and web applications to make conclusions from sequencing data. The clearest example of this is when potential novel SNPs have been found, and it isn't clear which represent real variants and which might simply be errors that may have been, for example, caused by incorrect base calling, or introduced during library preparation or during sequencing by synthesis. Similarly, beyond simple SNPs, larger-scale issues with genomic DNA can include problems with analysis of potential *indels*, insertions or deletions of stretches of nucleotides. Similar to putative SNPs, it can be difficult to be certain if an indel, such as a gap in a sequence or an unexpected series of nucleotides present where it should not normally be, represents an error in sample preparation or the sequencing reaction, or is a bona fide chromosomal anomaly. This is particularly problematic with massively parallel NGS techniques as these require combining the individual reads and aligning them along a standard reference sequence. Indels can be extremely hard to interpret as they won't match up appropriately with the reference sequence being employed for comparison.

One recent technological advance, particularly used in reversible dye terminator sequencing, involves bidirectional sequencing of each individual template fragment. This paired-end sequencing means that two sequencing-by-synthesis reactions are performed for each small piece of DNA in the library being sequenced. This greatly improves the efficiency of mapping—in particular, for indels. However, very long read techniques like nanopore sequencing might still be better for long repeating sequencings and larger indels.

Different NGS techniques demonstrate vastly divergent pros and cons, ranging from cost to throughput to rates of the different types of potential errors. Given these inherent differences, sequencing the same starting genome with more than one method will result in distinct sequences. Although the error rate of any sequencing technique might be quite low, especially with high degrees of coverage, when analyzing huge targets like a complete human genome, the results can be very hard to interpret.

When two sequences generated from the same starting genome have been produced with different methods and compared, large numbers of potential SNPs found in one sequence have actually not been observed in the other. In these types of comparative studies, the percentage of SNPs found in both sequences has ranged from approximately 90 percent to only about 65 percent. Furthermore, these types of analyses have shown that overall error can range from 0.1 to 0.6 percent, hence there can be a significant number of possible SNPs actually observed simply from error. Even in the best case, significant potential exists for misassignment of an error as a SNP, or potentially vice versa. Novel SNPs should always be verified through a secondary technique, such as a targeted approach—for example, traditional Sanger sequencing of the specific region of interest.

Another extremely useful method for WGS comparison is what is generally referred to as "trio" analysis. This means sequencing an individual and their two parents in combination. The human mutation rate is less than a hundred new changes per generation. Thus, the vast majority of your DNA is going to be largely identical to the sequences that can be found in the genome sequences of your parents. The information gained from trio sequencing can certainly be employed to validate potential SNPs and other mutations and issues found in WGS of an individual. However, trio sequencing is also being used to determine which regions of the genome seem more prone to generating errors with specific WGS techniques. This information could then be used to model error rates in a certain context, and potentially to develop data-processing filters to increase overall sequence quality.

Since most NGS techniques are fast and cheap but prone to error, one strategy that has emerged is performing NGS as a first step to extract a rough draft of a WGS, which is then analyzed by other more focused methods to validate novel SNPs or indels, especially those that might change the protein-coding sequences of specific genes, or otherwise cause significant functional changes. These specific variants of particular interest might then be resequenced in a targeted and directed way with a higher degree of accuracy, for example by Sanger sequencing with PCR primers specific to only that region of the gene.

It is natural to consider which of all the sequencing techniques might be best. But currently, the answer to this question seems to depend upon the specific application, the length of the target, and the amount of error that can be tolerated by the analytical goals.

Reversible dye terminator sequencing has emerged as the generally accepted standard for human WGS. This technique is extremely cost effective and fast. Excellent library preparation and data-processing tools have been developed for this platform. In fact, sample preparation is so straightforward that it does not take a great deal of skill or training.

SNPs, most indels, and shorter repeated regions can generally all be successfully identified with reversible dye terminator sequencing. However, the short read length of this method decreases the ability to correctly align long repeated sequences. That being said, this is less of an issue in practice, since we have good reference genomes to guide data processing and most relevant SNPs and other mutations won't usually be found in these hard-to-map regions. The reasons above also make reversible dye terminator sequencing an excellent choice for studies focused on the exome, the complete combination of all exons, the regions of the genome expressed as proteins. Furthermore, excellent kits for library preparation specifically developed for exome sequencing are available from biotechnology companies designed to feed directly into sequencing through this modality.

Exome sequencing has been emerging as an excellent choice, particularly for many clinical applications. Although the exome does not include most of the genome, as it does include the portions of genes that code for proteins, the exons, it will contain all potential coding mutations that would change protein structure. For basic research purposes WGS is still currently preferred, but exome sequencing may ultimately be better suited to most screening and diagnostic activities that require more information than can be obtained from analysis of SNP arrays—and may even be better suited for personalized medicine.

Exome library preparation kits that isolate and amplify all the exons present in a sample of genomic DNA have proven surprisingly flexible in extending beyond the species for which they have initially been designed. In particular, an exome library preparation kit developed for cows was used successfully for bison DNA. Similarly, researchers used a human exome library preparation to obtain the exome sequence of Neanderthals. It should be noted, however, that kits for exome library preparation must generally be developed for each new species analyzed and can be a significant stepping stone in studies of novel or rare species.

Nanopore sequencing is another sequencing technique that is gaining prominence for many applications. Although straightforward and fast, its incomparably long read lengths lead this technology to suffer from relatively high error rates. However, as only single-stranded DNA is fed through the nanopore channel, a recent advance in this technique has permitted both strands of the template to be read one after the other. This innovative development has reduced the error rate of this method from about 15 to 3 percent. Further reduction in error rates with nanopore sequencing has been observed by regularly shifting the voltage across the membrane in the device between two set values. This seems to result in alternatively stretching and compressing the DNA molecule in the pore, allowing for more information to be gained in the recorded current data. While an excellent advance in this particular technology, nanopore sequencing is still not well suited for routine WGS. There are many areas, though, where this approach is emerging as the technique of choice.

One potentially strong application for nanopore sequencing is epigenetics. This is generally defined as heritable mechanisms that control changes in an organism, excluding direct variation in nucleotide sequence via mutation. One primary aspect of epigenetics is the chemical modification of nucleotides in ways that ultimately modulate gene expression. In this case, it is not the DNA sequence that is important, but rather the chemical changes the nucleotides have undergone that make genes more or less likely to be expressed. Thus, accurate identification of the site of epigenetic changes to DNA is critical to many investigations into development, health, and disease. Nanopore sequencing can be used to identify modified nucleotides associated with epigenetic changes in gene expression, as these altered nucleotides will lead to predictable differences in the current measured through the pore as they pass.

Another area where nanopore sequencing excels is species identification and analysis. There are very small nanopore devices that could be readily deployed for use in the field for pathogen (disease-causing virus or bacteria) identification and epidemiological research that can be used to look for horizontal gene transfer events, such as when genes associated with human disease can be swapped between bacteria. Because you do not need extremely accurate DNA sequencing to identify species or even specific genes, for these types of applications the benefits of nanopore sequencing far outweigh the limitations generated by the relatively high error rate.

Nanopore sequencing can also be used to analyze mixed populations of organisms, for example, to characterize the species present in a particular sample, such as the microbiome present at a specific time and place, for example, the bacteria in an organism's gastrointestinal tract. However, these types of applications can require performing many reactions at the same time to increase throughput, a technique referred to as parallelization. There are larger nanopore sequencing devices that can support up to 144,000 channels in 48 individual flow cells with 3,000 channels per cell, each of which can read up to 450 nucleotides per second. In a single device, nanopore sequencing can theoretically generate a truly staggering amount of sequencing data. Up to fifteen terabytes of data could potentially be generated in a single sequencing run, which raises the extremely important question of what to do with all this data.

This is an extremely exciting time for DNA sequencing. Although Sanger sequencing has in many ways remained the gold standard for decades, many other techniques and technologies have recently emerged, each with different benefits and limitations. While there doesn't appear to be one ideal DNA sequencing modality that is best for each and every question and application, different niches have been developed that vary in length, accuracy, speed, and cost requirements. What remains constant among most of these specific methods, especially with WGS and exome sequencing, is the huge volume of data that must be processed, aligned, compared, analyzed, and stored. Therefore, in many ways, the real challenge of the postgenomic era might be the "big data" issues that are also emerging in many other fields.

The huge acceleration in DNA sequencing data collection shows no indication of slowing, and some of the most critical and important innovations ultimately hinge on developing new solutions for handling these truly astounding data sets. Fast fiber-optic data transfer and cloud computing are both rapidly emerging as essential to processing and analyzing genomic data. However, without significant computational and analytical advances in storage and data analysis algorithms, we may soon find ourselves with more data than we know what to do with.

CHAPTER 5

Big Data!

O ur ability to generate genomic data is increasing extremely rapidly through the proliferation of better and higher-throughput sequencing instruments. As advanced robotic DNA sequencing instruments have made the production of genomic data easier and more automated, figuring out what to do with all this data is fast becoming the major bottleneck to realizing new discoveries. The big data problem involves data transfer, processing, storage, and analysis and is leading to innovative hardware and software solutions.

There are many different specific techniques for generating genomic data, and while whole genome sequencing might seem the ideal for data collection, especially in the search for potential disease-causing mutations, if 99 percent of the genome is not made up of the exon regions that code for proteins, the exome might be a more efficient standard for some applications. Similarly, systematic determination of the presence of known variants in individuals and groups within different populations through SNP analyses focusing only on the specific patterns of polymorphisms present can also be very useful and significantly more straightforward than large-scale sequencing. That being said, the future that seems to be emerging actually involves comparative analyses of a lot of data—many whole genome sequences, up to hundreds of thousands at a time—in the search for the genetic basis of complex conditions such as cancer, diabetes, and cardiovascular disease.

Other scientific pursuits also generate huge amounts of data. For example, CERN (an atomic physics lab), the site of the Large Hadron Collider, can generate a petabyte of data per second. A petabyte is one million gigabytes, or a thousand terabytes. A DNA sequencer can generate a few terabytes of data per run, but each run can last several days. Although a particle accelerator may generate a much greater quantity of raw data than DNA sequencing, a huge amount of that data is filtered out and never actually saved, processed, or analyzed. Filtering and compression can be employed with genomic data, but it is extremely hard to know what needs to be maintained and what can be tossed out, especially in such a rapidly growing field, because the information that might not seem important today may end up being critical for future studies.

ARCHIVING DATA

When it comes to retaining data from sequencing, many questions arise. Do you need to keep every single raw read file from massively parallel next-generation sequencing? Do you only hold on to the final aligned genome sequence? Most genome sequence files also include scores for each nucleotide position estimating the overall confidence that the nucleotide at that site is correct and not a sequencing error. In some cases—for example, when comparing different sample preparation methods—a confidence value that represents a statistical measure of the potential for a base-calling error might be a critical part of your analysis. Is this really needed for all uses? What about a filtered file that only includes confidence values below some threshold level, so you know which nucleotide assignments might be questionable? And what if the cost of re-resequencing is less than that of long-term storage of large amounts of data?

It should also be noted that some funding bodies, such as the National Institutes of Health, require that researchers archive their data for several years after a project is completed. How exactly DNA sequence data is defined is more than just a semantic issue. This question gets deep into matters of scientific method and reproducibility. One key expectation in science is that conclusions should be validated by other groups doing the same experiment and getting the same answer if they use similar methods and tools. However, from library preparation to the specific DNA sequencing instrument selected to the data processing and analysis software application employed, there are many opportunities for two groups to reach different results in response to the same basic genomic question—for example, identifying the specific SNPs associated with a particular disease state. Thus, keeping sequence files can be an important resource so that others can look at them as questions and tools change over time. But should only the raw data be retained? Only the final aligned sequence? Only what was actually analyzed? Everything? Databases full of sequencing files can serve as the basis for investigations by researchers without access to the samples or technical resources required to generate their own raw data.

ACCESSING DATA GENERATED BY SEQUENCING

Although the exact size depends upon a number of factors, a single human genome sequence can take up about two hundred gigabytes of storage space. That is about the capacity of the hard drive in an average laptop. So if a DNA sequencer is generating one terabyte per day of raw data and needs to store all of it, it would quickly fill standard hard drives. Because of data-storage limitations, off-site server arrays that link dedicated sequencing facilities to individual researchers through high-speed connections are becoming more common. Furthermore, researchers are really interested in more than just single genomes. There are multiple large DNA sequencing projects worldwide, and if one wanted to compare data from five thousand human genomes, that would take up a petabyte of data, which must be stored, processed, and analyzed. Ranging from large-scale databases that store DNA sequences to the processing power required to actually perform analyses, genetics and genomic medicine are currently facing a crisis of inadequate computing infrastructure and expertise to handle all the information.

Because genomics data only exists as computer files, innovations in computer science, computational biology, bioinformatics, and statistics are proving essential for dealing with the big data problem. Several national and international groups with public and private funding have emerged to offer solutions for these issues. In the United States the National Center for Biotechnology Information has long maintained open-access web portals that permit anyone to upload sequence files and query various databases, whether for specific genetic variants or comparing multiple whole genome sequences. This has huge utility for genomics researchers to tackle questions ranging from rapid and simple determination of how similar two DNA sequences are to generating taxonomic trees of related species to illustrate evolutionary change over time. Although database searches and meta-analyses are huge areas of rapid development, at the level of individual sequence data, significant need exists for what might otherwise seem like very straightforward questions about limiting sequence determination and analysis.

As whole genome sequencing proliferates, one major issue that requires significant development is discriminating between sequencing errors and true variants. Much of why we sequence individual genomes is to conduct comparisons looking for mutations that might be associated with a particular trait or disease, but if you don't know whether a potential single nucleotide polymorphism (SNP) identified in a sequence file is real, the ability to make any further conclusions regarding the possible mutation that has been identified is greatly limited.

Although some potential variants can be validated through performing alternative genomic analyses, this step can be time consuming and might not be effective for all types of polymorphic differences, mutations, and chromosomal abnormalities. Developing better automated means for evaluating potential sequence variants and determining whether they simply reflect errors in sequencing would be extremely useful. As with any specific goal that includes computer scientists, determining the most efficient way to get the best answer is also critical. Any approaches that might be employed to analyze DNA sequence data will need to be efficiently applied to large data sets and would need to work as fast as possible and employ as little computing power as might be required to arrive at a reliable answer.

One perspective that could assist in analysis might be preprocessing in such a way as to simply highlight potential variants relative to a gold standard reference sequence. Instead of listing each and every nucleotide, a prefiltered file for analysis might only include the locations of variants. This would be like having a 3-D map of a number of similar haystacks and highlighting the locations of all the needles within each one. However, generating data with the ability to discriminate between true variants and different types of sequencing errors at sufficiently reliable levels can be a challenge.

One way of looking at this issue is to ask what differences might exist in the patterns underlying true SNPs compared to simple sequencing errors. DNA is made up of four different types of nucleotides. Cytosine (C) and thymine (T) are referred to as pyrimidines and include a six-member ring with a hexagonal molecular shape. Adenine (A) and guanine (G) are referred to as purines and include a six-member ring linked to a five-member ring—a hexagon stuck to a pentagon, each sharing one side. Pyrimidines, C and T, share a similar overall molecular shape, as do purines, A and G.

Mutations that mistakenly swap one pyrimidine for the other, a C for a T, or a T for a C, or alternatively one purine for the other, an A for a G, or a G for an A, are referred to as transitions. The other type of mutation, where a pyrimidine (C or T) is swapped for a purine (A or G), or vice versa, is called a transversion. Because of the molecular similarity between two pyrimidines, or two purines, transitions are much more frequent than transversions. By knowing something about the molecular basis for mutation, the probability that a particular potential SNP is real and not an error can depend upon whether it appears to be a transition or transversion. In short, transitions have a higher probability of being real SNPs, while transversions are more likely to be a sequencing error.

This is the kind of information being put into software that looks at potential SNPs and determines whether they are likely to be real or errors. Another consideration informing these types of analyses is the assumption that errors occur randomly within a sequence. Real SNPs, however, probably won't be seen as often within genes in positions that will change the amino acid sequence of the protein encoded because mutations that alter protein structure could have negative effects on the person with the mutation. Innocuous SNPs outside protein-coding regions would build up over time and be more likely to be observed when comparing an individual genome sequence to a reference standard. As errors can show up anywhere, potential SNPs within protein coding regions have a higher probability of just being errors. Thus, with information about where specific genes are located in the whole genome sequence and detailed knowledge of the structure and function of proteins that might be altered by potential mutations, that information can be used to inform software applications for distinguishing SNPs from sequencing errors.

An important guideline for comparing individual genome sequences to a reference standard is referred to as the Hardy-Weinberg principle. Basically, it states that although each generation will see the emergence of some new SNPs, unless the overall evolutionary pressures change, the population should not display significant incorporation of new mutations that would change protein structure and function. Sort of like the genetic version of "if it ain't broke, don't fix it."

The Hardy-Weinberg principle is really more of a statement of overall population genetics and how it might change over time than a specific way of looking at an individual genome. That being said, it does help determine whether potential SNPs that would change protein sequence in significant ways might actually instead reflect errors in sequencing. Knowing that a particular SNP is probably a real finding and not just the result of an error in sequencing is really just the beginning of the process, especially for studies looking for the genetic basis of disease.

DETERMINING IF A NOVEL SNP IS CLINICALLY RELEVANT

The inherent variability of the human species is actually quite low. There are currently almost eight billion people on earth.

Evolutionarily speaking, however, this huge population has resulted from very rapid expansion over a relatively short time. Some estimates suggest that in less than a hundred thousand years the human species has grown from a population of as few as ten thousand individuals. It probably wasn't until the advent of agriculture about ten thousand years ago that the human population rose above a million individuals. This rapid growth from relatively few ancestors has resulted in many people who are greatly similar to each other. Our genomes do not display the same level of diversity found in many wild organisms, including our closest evolutionary relatives, primates.

Genetically speaking, many nonhuman primates are extremely similar to humans but maintain much higher levels of genetic diversity within their species than humans. Orangutans, for example, have three times as much genetic diversity as modern humans. Looking at the specific amino acid sequences of all the proteins, the so-called proteome, chimpanzees are 99.4 percent identical to humans. Comparing protein sequences can be particularly informative as, given that most amino acids are encoded by more than one three-nucleotide codon, differences at the DNA level can be irrelevant when looking at the level of proteins. As long as two proteins have the same sequence, they will function identically, regardless of which specific codons were present in the genome encoding them. However, these differences at the level of DNA can affect gene expression. Looking at our close evolutionary relatives might be another way of determining whether SNPs found in human genome sequences might be potentially involved in causing disease, or might simply be benign or so-called neutral mutations, neither particularly beneficial nor harmful.

Comparing genomic sequence data from different nonhuman primates to find sites where SNPs are common will most likely reveal that these are not the cause of disease or otherwise decreased health relative to the rest of the population. Humans share a great deal of similarity to our recent evolutionary relatives, particularly in protein-coding regions of the genome. These SNP sites found in nonhuman primates can be logged and, when observed in human genomic data, can be excluded as potential causes of disease. While this type of approach involves very large data sets to develop and in the end only provides a statistical probability for any particular SNP, it is an excellent example of how innovative perspectives might answer critical questions moving forward, such as which SNPs are actually functionally relevant when found associated with a particular disease.

Studying genomic data is quite different from the types of experiments biologists typically perform. Traditionally, a great deal of basic scientific research took place in universities, and then developments might be handed off to pharmaceutical companies and possibly lead to clinical trials of a medication or treatment for a disease or disorder, performed by medical doctors and nurses. At the center of it all, individual research teams would address a specific question employing specialized techniques.

One common theme throughout modern pharmaceutic research has been the search for enzyme inhibitors. Ranging from infectious disease to cancer to depression, if you have a specific protein (such as an enzyme) that you know is important in a disease process, you can study its structure and function and then design or otherwise identify chemicals that might block the function of the enzyme. This is like trying to find the right key to a lock.

Although there have been transformative technological breakthroughs in biomedical research that have facilitated these traditional approaches, the rapid growth of DNA sequencing data is truly unlike anything previously encountered. While interdisciplinary teams consisting of medical doctors, genomics researchers, computer scientists, and other specialists are working together to develop tools and techniques for large-scale genomic analyses, the wider big data problem in genomics is requiring much larger partnerships between governments, industry, and academic researchers.

INNOVATIONS IN DATA COLLECTION AND ANALYSIS

The New York Genome Center and the IBM Watson Group have been working together for approximately five years in an effort to apply artificial intelligence (AI) to genomics data—for example, to develop a deeper appreciation of the genetic mutations involved in various forms of cancer. Personal genetic testing companies like 23andMe are partnering with pharmaceutical companies in attempts to better develop personalized treatments that depend upon the unique genomes of patients. The pharmaceutical company GlaxoSmithKline recently announced a $300 million investment in 23andMe that will support research in innovative pipelines for new drug development employing the huge genomic database that 23andMe has collected.

The major direct-to-consumer personal genetic testing companies are well placed to maintain their market dominance and resist advances from smaller rival start-ups. As they have the majority of the existing market share, they have developed inroads that permit working together with pharmaceutical companies. Because a limited number of people will be interested in personal genetic testing and most individuals likely won't be tested by multiple companies, this benefits the already-dominant players. The so-called network effect means that, similar to what happens with a competitor to Facebook or Twitter, already having a significant user base can protect a company's position, even if a newer rival rolls out innovative advances.

However, it is not only genomic data that needs to be collected and analyzed. To assist in overall determination of the genetic basis of disease and complex traits, clinically relevant data ranging from information like height, weight, and age, along with more complex data such as blood pressure and cholesterol levels, also need to be obtained, stored, and analyzed. There is clearly also a role for publicly funded activities in this area. The UK Biobank, for example, is currently collecting DNA sequence data and detailed medical information of about half a million individuals. However, this data collection and storage does, of course, lead to some complex legal and ethical questions concerning privacy and the potential for financial exploitation, especially as much of this information collection crosses international borders. Furthermore, the vast majority of DNA sequence data that has been collected has been from people with white European ancestry. This homogeneous sampling is troubling as it may limit the applicability of any significant findings to the broader human species.

Biomedical researchers have historically focused on developing specific hypotheses based upon previous information, and then, following experimental testing, a likely mechanism was considered to explain a particular end point. For example, if a tumor in a specific tissue responded to a particular drug, that same strategy might be tested to treat a similar type of tumor found in a different tissue.

This is all very different from the emerging big data perspective where we collect as much information as possible and then look for statistically significant correlations, letting the data develop the hypothesis automatically. Thus, the big data approach focuses on the use of what are referred to as genetic biomarkers. What this means in practice is that it is possible to compare the mutations found in DNA sequences derived from many similar tumors originating across a variety of tissues from a huge number of individuals, and if the common causative mechanisms are found, then therapies could be developed to treat tumors according to their observed molecular signatures, rather than empirically according to trial and error and best guesses. As some estimates suggest that ultimately only about 10 percent of drugs in clinical trials ever prove useful, a new paradigm for drug discovery that focuses on genomic personalized medicine is certainly appealing.

Working from big data requires the development and application of new types of machine learning and AI approaches. It also means that the way biomedical research scientists have traditionally been trained to work may not be a good fit for the current needs in this area. To date, most medical genetics research has focused on the search for extremely rare mutations in single genes that cause recessive diseases such as cystic fibrosis. The lessons from this perspective might be too restrictive moving forward as scientists attempt to understand complex diseases that involve contributions from many different subtle genetic variants, as well as behavior and environment. The historical research paradigm has had an influence on how labs are set up—most research laboratories have allocated much more wet-lab bench space where experiments are conducted than space for computers and data analysis. It seems as though rethinking the entire research enterprise—from our educational system to architecture and design—may be required to better support new approaches to research. Are we moving toward a reality where automated sequencing facilities flow data directly to supercomputers for analysis, without any direct human engagement?

The big data era of genomics may require more computer science than actual biology. Sample preparation, data generation, and analysis were all done manually, and now all these steps are becoming more automated. Introduction of techniques like machine learning and AI into genomic research might make the whole process completely automated. We are approaching an era when an individual can provide a sample (usually saliva) through the mail to a genetic testing company like 23andMe and then robots can process the sample. The DNA sequencing instrument will then generate computer files, which can be automatically analyzed through self-improving AI that generates the answers to questions the individual researcher didn't even know to ask, such as the specific pattern of SNPs associated with a particular disease, ancestry, or even something like the preference for vanilla or chocolate ice cream. In this brave new world, could it be that only the highest levels of researchers and clinicians will have a required role?

For all this potential for revolution and disruption, genomics hasn't yet become the panacea to explain all complex human diseases and introduce novel preventive and therapeutic interventions to rid our world of heart disease and cancer, the leading causes of death in the United States. This is, of course, early days. However, going forward, managing expectations and reducing sensationalism is critical. Although predictive tools for understanding the genetic basis for complex human diseases and other traits are emerging, in many cases these involve huge amounts of data to develop, and can usually only be applied to relatively homogeneous groups of people. Additionally, any group of variants associated with a particular trait or risk measurement will usually only explain a small percentage of the total variance within a population. For example, if your particular genes put you at a lower risk for high cholesterol, but you eat cheeseburgers and fries every day, you probably will still end up sick from that diet.

Another significant risk is rushing to explain everything through genetics. This can sound similar to the disastrous results of eugenics research a hundred years ago, where top researchers fueled misguided xenophobic and cruel social perspectives, attempting to rationalize racist and prejudicial political agendas through supposedly scientific means that were at best tenuous and at worst indefensibly evil. Nobody is currently endorsing forced sterilization of "undesirables," but the present climate does present a context where the suggestion that genetic forces inexorably determine different potential outcomes might be a real concern.

That being said, all this research and innovation is happening now, whether we are ready for it or not. In vitro fertilization (IVF) centers are using genetic screening to assist would-be parents in selecting embryos for implantation. Population genetics is employing studies of hundreds of thousands of individuals. Over 26 million people have taken personal genetic tests from companies like 23andMe and AncestryDNA. This data is being used for everything from understanding human evolutionary biology to quantifying individual disease risk.

Advances in genomics require interdisciplinary collaborations. Although it makes sense to imagine geneticists and clinicians working together to further our understanding of the genetic basis of disease, the techniques and technologies involved can be impenetrably complex to all but a select few experts. Furthermore, the computational and statistical knowledge required to make definitive conclusions through analysis of genomics data represents a very high bar indeed for the average biologist or physician. Even if the right questions are asked, the proper procedures followed, and the appropriate controls performed, the correct computational and statistical expertise must be available, or else incorrect or misguided conclusions can be made. When these studies have the potential to lead directly into clinical medicine, ensuring that proper procedures were carried out is particularly essential. However, genomic medicine isn't the only area where incorrect analyses can lead to potentially bad outcomes. Another major area with a great power to improve understanding, and a significant potential risk of improper interpretation, is the use of personal genetic testing to analyze ancestry.

PART III

Applications for Genomic Information

CHAPTER 6

Ancestry

DNA tests can be very accurate when there is a focused ques tion to be addressed, such as determining whether an individual carries a specific gene variant (e.g., the deletion of the 508th amino acid in the sequence of the cystic fibrosis transmembrane conductance regulator, otherwise known as *CFTR* ΔF508). Someone who inherits two copies of *CFTR* ΔF508 will develop cystic fibrosis, so most prenatal genetic testing specifically assesses whether a prospective parent is a carrier of this mutation. Similarly, comparisons of different genomic DNA sequences—or, more accurately, patterns of inheritance of genetic markers—can very easily be used to identify relatives, such as testing for paternity. Personal genetic testing works in a similar fashion. However, rather than comparing specific gene sequences between specific individuals, or looking to identify a particular gene variant, these companies compare an individual pattern of single nucleotide polymorphisms (SNPs) to reference populations in large databases employing complex analytic methods. These personal genetic testing reports will often include information about an individual's family heritage, generally the geographical regions where their ancestors may have come from.

SNP PATTERNS AND ANCESTRY

What does a SNP pattern mean and how does it factor into identifying an ancestral region from a DNA sample? AncestryDNA, for example, looks at over 700,000 different SNPs—that is, over 700,000 different locations in the human genome where there are known to be SNPs that can vary depending upon the genealogical ancestry of that individual. DNA is analyzed by seeing which nucleotide is present at each of these locations. The analysis performed by these services is not done through genome sequencing but by having a collection of the different small regions surrounding the SNP locations put into a format where each specific SNP can be assessed using a technique referred to as a SNP array.

As discussed, DNA is double stranded, and the two strands bind together in a predictable fashion. So if a small piece of DNA corresponds to a region around one particular SNP variant, after isolation of genomic DNA from the sample provided, it can be determined whether the sequence at this site is complementary to that small SNP-containing piece of DNA in the test. If it is, then it is positive for that particular SNP; if it is not, then it doesn't carry that specific SNP sequence. This analysis of nucleotides is done for all locations being tested using a SNP array, and then the resulting SNP patterns are identified and evaluated using computer programs that compare them to known patterns in reference databases.

A SNP array is performed with many tiny samples of single-stranded DNA, each containing one particular SNP-containing sequence of DNA previously identified within the genome of a specific population. A SNP array is basically a small slide that has a regular grid on it onto which the small DNA sequences containing known SNPs are deposited. Small pieces of single-stranded genomic DNA from the sample being tested are then added to the SNP array on the slide, and those pieces of single-stranded genomic DNA from the sample being tested that are a complementary match to the ones on the SNP array will bind together to form small pieces of double-stranded DNA. A machine then reads out which spots were bound, as each has a known location on the SNP array. This process demonstrates which of the different SNPs is in an individual's genome by identifying the specific sequence at that location. This is basic but incredibly powerful considering the huge number of SNPs assessed, and the large database of previously analyzed DNA samples with known SNP patterns employed for comparisons. Previous analyses of specific populations have found particular corresponding SNP patterns within those historically distinct groups of people. So if a DNA sample sent to AncestryDNA or 23andMe contains SNPs in that same arrangement, it's likely there is some ancestry from that group.

To get a sense of how this works, imagine that you have a group of people who are all part of a sports team. Although you know the height and weight of each team member, you don't know what type of team it is. Is it a basketball, baseball, or football team? If reference data sets exist for all the members of all the NBA basketball teams, MLB baseball teams, and NFL football teams, then by cross-referencing the values for the team in question with a database of all types of teams, you could figure out what type of team you have.

I am six foot three and weigh about 200 pounds. I could be a point guard, a kicker, or maybe a quarterback in football, and any of several positions on a baseball team. Thus, by only knowing one team member's height and weight (or a particular individual SNP in the genome), you can't identify to which type of team (or genealogy) that individual likely belongs. However, if you know the pattern for the whole group, you can. If the statistics show a lot of tall and thin people, you probably have a basketball team. If there are a few smaller people, but most people are heavy for their height, and some are tall and heavy, you probably have a football team. Similarly, if the group of people was a mix between basketball players and football players, you might be able to figure that out, too, if some are very tall and thin and some others are heavy and somewhat tall. Clearly, there are some scenarios in which this would not work: if there were no recognizable pattern, if the person's SNPs were a mix of too many populations, or if you only had one player from each different type of team in the reference data. The size of the data set matters—if people in the group in question were members of a very rare type of team that wasn't in your database, say rugby or jai alai, you would be hard pressed to correctly identify their origins.

AncestryDNA looks at over 700,000 different SNPs and has an ever-growing database with millions of reference genomes, along with powerful analytic and statistical tools. According to AncestryDNA, information for over 15 million people is present in their databases. With all that data, they are likely to be able to give you a sense of your ancestry. But this raises the question, What does it really mean if a genetic ancestry report says you are x percent something (country/ethnicity)?

WHERE ARE YOU FROM?

To understand genetic ancestry reports, we must think through a bit of human history. *Homo sapiens* evolved in Africa, and then there were periods when populations migrated away, ultimately spreading throughout the world. Broadly speaking, some of these groups settled in a particular location and then remained there, neither moving on nor significantly mixing with others. This reproductive isolation is the key to understanding ancestry.

For example, the Japanese are a relatively genetically homogeneous people. There have been permanent residents in Japan for thousands and thousands of years, and compared to some other cultures, there has been little interbreeding with outsiders. Similarly, Ashkenazi Jews are generally genetically distinct from the populations that surrounded them in Europe for hundreds of years. Because of this general lack of genetic diversity, there are many genetic diseases found within the Ashkenazi Jewish population, and there are specific prenatal screens and even personal genetic tests for people with Ashkenazi heritage. One example, among many, is Tay-Sachs disease, which is a rare neurological disease much more common in Ashkenazi than the general population.

Humans have been in the Americas for around fifteen thousand years, but for the vast majority of that time there were no people of European descent here. However, genetic ancestry and cultural identity sometimes need to be looked at from very different points of view. When Elizabeth Warren released results of a genetic test apparently demonstrating some Native American ancestry, this prompted sharp commentary—and not only from her political rivals. Representatives of some Native American communities pointed out that membership in a tribe or Nation involves a great deal more than distant bloodline ancestry, leading Warren to apologize to a representative of the Cherokee Nation. One reason that this could have been a particularly sensitive issue is that Native American interbreeding with white Europeans might not always have been voluntary. This cultural-genetic dichotomy is not unique to Native Americans. Inclusion and exclusion within a particular group can exist according to social and political agendas rather than solely according to biological origins.

The maps of countries in the Americas, Europe, Asia, and Africa that we know today wouldn't mean much to our distant ancestors. Borders drawn for sociopolitical and socioeconomic reasons don't often directly correspond with genetic distinctions. While a genetic test might be able to tell you if you had a recent ancestor (such as a great-grandparent) who was white European when you broadly consider yourself black African American, distinguishing between English and Irish can be an entirely different question, especially when you consider that Northern Ireland has only been in existence for about a hundred years.

The answers we seek from genetic testing are only known from the analyses that have been performed on the data that has come before, and how they compare to our own results. Whether looking at ancestry or the genetic basis of disease, past correlations can be employed to make conclusions, create screens, and lead to further hypotheses to test. Because people from specific populations share genetic information that is distinct from other populations, those differences in specific SNP patterns are then regarded as markers for that particular ancestry. However, whether this is definitive information all depends upon how you qualify the determination you are trying to make, how much genetic information you have collected, how different it is from other groups, and the power and accuracy of the data analysis and statistical tests employed to make the determination. Specific ancestry can be a moving target, and analyzing the same genome in different ways, especially as databases and algorithms evolve and grow, can lead to distinct and even seemingly incompatible results. Sending samples to two different genetic testing companies could lead to confusingly different conclusions if the two companies are using different databases and algorithms.

Even if you know you have ancestors from a specific area, you might not get the results you are expecting. Their ancestors might not have actually originated from that region. Thus, their DNA markers might reflect the earlier location of origin, especially if they emigrated with a relatively closed expatriate community. Also, you might not show the presence of a certain ancestry even when your forebears did all come from a specific area if you didn't end up inheriting any of those specific SNP markers. The probabilities behind this type of analysis get very complicated very fast. In essence, if a person from a distinct genetic background breeds with someone from another group, and then the successive generations don't breed with the original population, it is certainly possible that a personal genetic ancestry test won't show the presence of any markers specific to that original group.

Although 700,000 SNPs might seem like a lot, it all depends on how many are distinct between populations, and how many generations of mixing have occurred. According to AncestryDNA, less than 1 percent of your DNA should have been inherited from any individual ancestor from seven generations back. Assuming something like thirty years or so per generation, this means that only about 1 percent of your DNA comes from any ancestor that lived as recently as two hundred years ago. Inheritance of specific SNPs is also subject to the fact that if you have one variant on one chromosome, and another on the other chromosome, it is a matter of random fifty-fifty chance which of these gets passed to the next generation. Thus, there are a number of reasons why your DNA test might not show that you are significantly Italian, even if you have always heard that your family has some ancestral roots in Italy.

One reason it might be difficult for certain people—for example, Americans of white European descent—to get truly meaningful ancestry information from personal genetic tests is that America is, globally speaking, a very new country spanning a geographically large area and has a very diverse population compared to some other countries. To some extent, this inability to target ancestry information stems from the presence of individuals of mixed descent and the dizzying array of heritages, cultures, and languages often living in close proximity in the United States. Often, the best answers one can get suggest that a person might be x percent Eastern European, or y percent South Asian, or z percent sub-Saharan African. These results arise from a complicated automated computational analysis of the different SNP patterns found in a DNA sample. Given the amount of information currently present in the reference database, more specific analyses might not currently be possible. It all comes down to how many people have been previously analyzed and placed into the database that have a similar, relatively homogeneous genetic background—and are distinct from other groups. If you are hoping a personal genetic test will tell you which village in Wales your ancestors came from, you will most likely be disappointed. Rather, it might say that you are English, or even just generally Western European.

There are many places in the world where national borders have been superimposed over very long-standing populations. Examples such as the partition of India and Pakistan emphasize how some distinctions—for example, if you are Hindu or Muslim—can matter much more than seemingly arbitrary geographic borders based upon political compromises. People's identities and ancestries don't necessarily correspond to easily determined genomic signatures. It all comes down to whether your ancestors lived in populations that were relatively reproductively isolated, and the extent to which those groups are well represented in whatever database is being searched.

The Alps are another example of how borders don't definitively show ancestry. In less time than it takes many Americans to commute into work, you can visit the villages on the French, Swiss, and Italian portions of the Alps. Similarly, even the cleverest analytical tools might have trouble deciding whether someone whose family has been living for generations on one side of the Rhine River in Alsace is of German or French heritage. That being said, with enough reference information regarding people from any geographically and reproductively isolated population, specific aspects of heritage could be identified within an individual.

RACE AND ANCESTRY

Genomic ancestry information has the potential to affect not only the way we see ourselves but also how we see others. Unfortunately, applying personal genetic testing to reinforce racist ideologies is already occurring. The question of racial identity is an extremely sensitive subject. America, as they say, is a land of immigrants. If your family arrived here two hundred years ago, and my family arrived here five years ago, as long as in our hearts and minds we are truly American, how can the difference in how long our families have lived here, which are minuscule to the point of being irrelevant on a genetic scale, make any true difference? Saying that only people of white European heritage are truly American is equivalent to white supremacy. Of course, "white privilege" is a very real thing and if you are subjected to racism because of how you look or your perceived ethnicity, then being nonwhite could be defined through the experience of exclusion from those that are viewed, or view themselves, as white.

From a biological perspective, what does "white" mean? It can't simply be determined by the amount of melanin pigment in your skin. This complexity and confusion exists because, when you get down to the details, race is broadly a cultural construct. Race, ancestry, and heritage are not synonymous. The way a person looks does not equal family, ethnic, or cultural heritage. Making assumptions based on appearance simply boils down to prejudice.

In many cases, at least when it comes to documentation and demographic data analysis, race is defined through self-identification. You put down "white" because you look white and your parents look white, and you were raised to define yourself that way. However, the advent of personal genetic testing is changing this for many people. Interbreeding between people of white European ancestry and those whose families trace their roots to Africa has certainly occurred throughout the history of our country—although in many historical cases procreation might have been without consent, and thus people may not have been able to openly discuss the traumatic true parentage of the resultant child, losing this information from the family history. Furthermore, descendants of African slaves brought to America may not have any detailed family history to begin with, including the times both before and during slavery.

If someone looks to be of white European heritage, but after taking a personal genetic test and doing a bit of online research finds out that he had an African American great-grandparent, does that change who he is? If a baby born to parents of Chinese descent is raised in America by adopted parents of white European descent, and never knows anything besides American culture and the English language, does that truly make her any less American than anybody else? Of course not. Being American isn't written in your genome, especially considering the relatively young and genetically diverse history of America.

To many Westerners, it can be difficult to differentiate among people from different Asian countries. Yet to people from China, Japan, or Korea, those national identities are incredibly strong and meaningful, for both historical and cultural reasons. African Americans and persons from Cuba and sub-Saharan Africa can all at first glance appear to look very similar to one another, at least to some people, but they come from diverse cultures, with different languages, and can have distinct patterns of genomic SNP markers. It should also be noted that studies regarding the genetic basis of disease are demonstrating significant differences in these populations, regardless of how similar individuals might "look" to some.

People of white European ancestry whose families settled in different regions of America feel those regional differences in many cultural and social ways. Someone from New York City who travels abroad may be confronted with the assumption that Americans are essentially Texas cowboys, but then this same person can stand out like a sore thumb as a "Yankee" when venturing into the Deep South.

These seemingly insurmountable differences are the result of a tiny amount of time on the scale of human history, not to mention human evolution. The genetic homogeneity found within truly inbred populations can be hard to come by in today's world—somewhat thankfully so considering the resulting significant incidence of genetic disease found in these groups. However, markers for the different previously reproductively isolated populations from which our ancestors emerged are still written in our genomes.

It is completely natural to want to belong to a group. Maybe this comes from the increasing isolation and loneliness people seem to be feeling. Similarly, the tendency to want to have one's own group seen as distinct or better than another group can be a natural progression from this need for belonging. The irony, of course, is that distinctions that from the outside can appear irrelevant can seem critical to the individual. If you come from Manhattan, whether you grew up on the East or West Side might be central to your sense of identity. If you live in Glasgow, it can be nearly a matter of life or death whether you support the Rangers or Celtic football (soccer) clubs. However, to most of the world, this distinction might seem ridiculous.

Sephardic or Ashkenazi, Sunni or Shia, Catholic or Protestant—these differences can be critical to individuals and can represent distinct geographical, cultural, and even genetic bases, but from the outside, it is very easy to simply see a Jew, a Muslim, or a Christian. A striking example that perfectly illustrates the sheer ignorance at the heart of cultural and racial prejudice happens all too frequently when Sikhs ironically find themselves subjected to Islamophobic abuse, just because they have brown skin and wear turbans.

Trying to integrate genetic genealogy and ancestry with our understanding of race and heritage is to a great extent like comparing apples and oranges. Furthermore, in many cases, the data employed in demographic analyses originate via self-identification, which may be neither accurate nor reliable.

DOES RACE HAVE ANY BIOLOGICAL BASIS?

How has our perception of race developed over time and how much of that change is because of actual genetic change or understanding? How do mixed-race people fit into this under-standing? What does it even mean to be "mixed-race" in light of the findings of genetic genealogy that almost everyone is to some extent a mixture of different populations?

The terms we use to describe race, ancestry, genealogy, heritage and culture are all jumbled together. What is the biological basis for the term "non-Hispanic white" found on many official documents? Doesn't this imply an existence of "Hispanic white"? What would that even mean?

Before they speak in their native tongues, can you tell the difference between someone from Portugal and someone from Spain? How significant is it to the understanding of the term *Hispanic* if a person comes from Central or South America, compared to if they call Spain home? If someone comes from Central or South America but "looks" Spanish owing to reduced shared ancestry with indigenous "pre-Columbian" populations relative to other people from the same country or region, does that make that person whiter? Again, what does that mean, and is it even a valid question to ask? What does that make people who might share more obvious characteristics with people indigenous to the Americas?

The Moors interbred with southern Italians and Andalusians, but in America currently we don't consider Sicilians any different from other Italians. It should be remembered that there was a time in America when Irish, Italian, and other immigrant groups like Eastern European Jews were not considered to be truly white. If a strict definition of white were followed that only included families of Christian descent with northern and western European ancestors that never mixed with any other groups, this would be tremendously limited and narrow minded—in short, racist.

So the question arises again: What does "white" really mean? As the saying goes, those in power get to make the rules, thus the repugnant antediluvian perspective that "one drop" of "black" blood, whatever that means, makes an individual *irredeemably* nonwhite. As different immigrant groups gain prominence and station, they can come to be seen as white. This is generally preceded by the acceptance of intermarriage and resulting offspring as white. However, the vestiges of slavery would seem to make this all but unattainable for African Americans, at least beyond the tenuous concept, socially and especially biologically, of "passing" for white.

A great deal of press, and no small amount of social and political turmoil, has recently centered on projections of the apparent decline of the percentage of Americans defined as white. The date when the country may no longer be majority white is feared by some and eagerly anticipated by others. Reactionary policies aimed at preventing this from occurring seem predicated on, at best, a definition of race that is not entirely biologically relevant. Furthermore, some individuals who could be defined as "progressive" seem to be operating under an assumption that once this threshold is reached, a multicultural utopia free from the influence of racism will emerge. However, evidence from some locations where different nonwhite populations live together in ostensibly constant strife might call this perspective into question.

PERSONAL GENETIC TESTING

I was raised Jewish and was always led to believe that all my relatives and ancestors as far back as anyone in my family knows have been Jewish. My AncestryDNA report states that I am 98 percent European Jewish. That is about as definitive as it gets. (Of course, my mother's knee-jerk response to hearing this was to reply, "That's it?") Furthermore, AncestryDNA's online system automatically matched me to my father and one of my sons, who had both previously used their services.

But what if my AncestryDNA test came back and shattered my belief regarding my Jewish heritage? What if through follow-up questions with my parents, I found out that I had been adopted from non-Jewish biological parents? The revelation of my true parentage would probably be more significant to my identity than questions of my Jewish heritage. Of course, Judaism is a religion, and religion is not an inherently genetic thing. In fact, from one point of view, anything that seems to suggest the existence of a "Jewish race" can echo past horrors such as the Holocaust.

If I had found out I was adopted, would it sting any less to find out that my biological parents were "at least" Jewish? What if I found out I was 50 percent Jewish because one of my parents was adopted? Would that matter?

The point of all of these questions is whether investigations into ancestry are done for informative purposes or to validate one's personal self-identity. How do we define ourselves based upon questions of family, race, and cultural heritage?

But this is not always simply a personal quest. The use of genetic testing to determine ancestry is also being applied in social and political contexts. People immigrating into Israel from former Soviet countries where evidence of Jewish ancestry may not be available are in some cases reportedly being subjected to DNA analysis to prove their right to Israeli citizenship. This is potentially a concern because, as with many analyses employed in genetic medicine, although the test being employed has statistical value overall, it can't necessarily serve a definitive binary sorting tool for all individuals.

On a more personal basis, genetic testing can pave the way for exposing family secrets. What happens if you and your brother take tests, compare results, and determine that you aren't actually full siblings? What are the ethical implications if through some genealogical sleuthing you determine that your biological father isn't really who you think he is? Are you compelled to confront your family with this information? As we will see below, the significance of these questions only increases in potential complexity and impact when medical issues are taken into account. As with any genomic data, the information contained within a personal genetic test depends upon the context and the perspective through which it is analyzed and considered. The questions being asked are as important as the answers.

Genetic genealogy can be incredibly useful and informative. It can also lead to shocking revelations. Building family trees through combinations of personal genetic tests, public records databases, and other sources of information can be very interesting and lead to personal connections, increased sense of belonging and community, and answers to long-standing questions about families and individuals. This same type of analysis can be used for other purposes, such as criminal investigations. Moreover, the data generated through personal genetic testing can be extremely valuable to biomedical researchers and the pharmaceutical industry. Another interesting aspect of some personal genetic testing relates to how much Neanderthal DNA an individual has. What does this mean, and how relevant is it to our understanding of ourselves as individuals and members of the human species?

CHAPTER 7

Neanderthal DNA

Neanderthals were close relatives of humans that for a time coexisted with ancient *Homo sapiens* but are now extinct. It is not exactly clear when and where the last Neanderthals died out, but the most recent remains that have been found are probably about forty thousand years old. The skeletal remains of many Neanderthals have been studied, and DNA samples have been successfully extracted and analyzed. In fact, whole reference Neanderthal genome sequences have been obtained, and these samples provide great insights not only into the genetics of alternative evolutionary lines that coexisted with early humans but also into our history, as well as some of the genetic basis for certain human traits—for example, particular types of hair and skin.

One groundbreaking report that detailed observations made from Neanderthal genome sequencing covered issues ranging from evolution to sexual behavior. It is known that hominins (humans, their direct ancestors, and close relatives, excluding apes) split off evolutionarily from chimps sometime around seven million years ago. This paper suggested that, extrapolating from the estimated mutation rate of the human genome, ancient humans and other hominins such as Neanderthals separated from each other evolutionarily around five or six hundred thousand years ago. In this study, genomes from Neanderthals and other hominins were analyzed from samples dated between fifty thousand and seventy thousand years old that were obtained from sites including caves in the Altai Mountains that span the western edge of Mongolia and the Caucasus Mountains at the southwestern border of Russia. Several interesting findings were obtained through this analysis.

The genome sequence of one female Neanderthal demonstrated that her parents were likely half siblings. In at least this case, close inbreeding occurred. Full siblings share 50 percent of their DNA, on average, and thus half siblings share 25 percent. Researchers could tell that her parents were closely related because, comparing the sequences of each of her paired chromosomes, each chromosome in a pair that was inherited from one parent was much more similar to the one inherited from her other parent than would be possible if her parents were not related.

There is also significant evidence that Neanderthals mated with humans and other extinct hominins. Humans likely also interbred with other extinct hominins. We know this because people from certain areas of East Asia and Oceania—in particular, islands such as New Guinea—contain DNA sequences also found in the genomes of the so-called Denisovans, another group of hominins. Interestingly there is evidence that human-Neanderthal hybrids may have had reduced fertility, but in other ways this interbreeding seems to have had benefits to the evolution of our species by introducing traits with potential selective advantages such as dermatological and immunological characteristics particularly suited to the environment outside Africa.

It seems that most of this human-Neanderthal mating occurred following migration out of Africa. Neanderthals left Africa long before humans, and although it is not clear exactly when and where humans and Neanderthals began to interbreed, the genome of a human that lived approximately forty-five thousand years ago already contains significant Neanderthal DNA. This DNA came from a thigh bone discovered in Siberia and represents one of the oldest otherwise modern human genomes ever sequenced. We still retain vestiges of Neanderthal-human interbreeding in our genomes. In one study, East Asians showed the most Neanderthal DNA (~1.4 percent), followed by Europeans (~1.2 percent), and finally Africans had by far the least (only ~0.2 percent). However, exactly where Neanderthal DNA resides in the human genome varies, and it has been suggested that altogether around 40 percent of the Neanderthal genome still resides within humans overall, when it is all added together across all humans analyzed.

By looking at the regions in the genome where humans had the greatest concentration of Neanderthal DNA, it appears to be preferentially found at sites known to regulate the protein keratin, which is involved in forming skin and hair. It was suggested that these Neanderthal contributions might have helped early humans physically adapt to conditions outside Africa. However, the presence of Neanderthal DNA can also be associated with a type of keratosis, a specific skin disorder that occurs following exposure to too much sun. Furthermore, Neanderthal DNA seems to affect our immune systems, in good and bad ways. Some Neanderthal DNA in the human genome might promote immune system function by protection against certain viruses, specifically those that Neanderthals had to cope with before human contact. Human-Neanderthal hybrids might have inherited these protective factors, which further promoted human survival in their new environments. However, Neanderthal DNA has also been found to occur at regions thought to be involved in autoimmune diseases such as Crohn's disease and lupus.

Neanderthals apparently had different brain shapes than humans; ours are generally rounder, while theirs were more elongated. Although there aren't any Neanderthal brains left to test directly, braincases can be used to estimate brain anatomy. In one recent study, specific SNPs inherited from Neanderthals were associated with brain shape in humans. People who possessed these Neanderthal SNPs had more elongated brains. Furthermore, these SNPs were also associated with increased expression of the gene *UBR4*, which seems to be involved in neurogenesis (basically, neuron cell proliferation), and decreased expression of the gene *PHLPP1*, which is believed to reduce formation of the so-called myelin sheath that surrounds neurons and aids with transmitting signals along cells in the brain. Decreased expression of the negative regulator of myelination could make signal transduction in the brain more efficient, but this was not directly demonstrated in this study. While this is not the first study to look at a role of Neanderthal DNA in shaping the human brain, it does go very deep into potential molecular mechanisms that might have led to changes in human-Neanderthal hybrids. To be clear, this doesn't necessarily demonstrate an influence of Neanderthal DNA on brain function, and to date no direct evidence has been demonstrated beyond an overall difference in brain shape between Neanderthals and humans.

There are over a hundred thousand different SNPs in the human genome with a Neanderthal variant—that is, sequences associated with the Neanderthal genome that most likely made their way into humans through interbreeding. 23andMe tests for 2,872 Neanderthal variants at 1,436 marker sites. You can either have a maternal or paternal Neanderthal variant at each marker site, neither, or both. Thus, for each marker size you can have zero, one, or two Neanderthal variants.

23ANDME, NEANDERTHALS, AND ME

The most Neanderthal variants any 23andMe user ever had was fewer than four hundred. I had a total of 259. This was apparently less than 72 percent of other 23andMe customers; 23andMe also provides information about a few human traits associated with some of these Neanderthal markers. In my case, one of the Neanderthal variants I inherited is associated with having less back hair than the general population. Of course, as this is only a statistical association, there is not necessarily a direct causative link between the particular DNA sequence of that Neanderthal variant and any known protein function directly related to back hair.

Another trait associated with a Neanderthal variant is having straight hair. Although I have straight hair, I don't carry this variant. This mismatch of traits and reality demonstrates the limits of these types of associative analyses. It might be hard for me to tell if I really do have less back hair than the general population, but it certainly is the case that you can possess a variant associated with a trait and yet not display that trait because of other genetic factors. Furthermore, you can obviously have a trait but not carry a variant that can be associated with that particular characteristic.

These traits are associations and should only be looked at from the point of view of probabilities, not certainties.

Look at it this way: if you drive a car with a seatbelt on, you are less likely to be injured in a crash. Similarly, if you don't wear a seatbelt the odds are higher that you will be injured in an accident. That being said, you can still be injured in a crash even if you wear a seatbelt; and similarly, although I wouldn't recommend it, if you don't wear a seatbelt and you get into an accident, you might not get injured. This is the way probabilities and associations work, and these examples demonstrate how they are different from direct causation, such as when a specific mutation that significantly inhibits the function of a specific protein causes a monogenic disease like sickle cell anemia or cystic fibrosis.

The genetic basis for complex human diseases generally involves many different genes, each usually contributing a small amount of the overall risk. But environment and behavior also usually play a role in whether a specific individual will develop one of these multigenic diseases (examples include cancer or heart disease). Ranging from broad statistical associations to testing for specific mutations, personal genetic testing as well as analytical and clinical genomics are paving the way toward a future where we can understand a great deal about our personal genetics, our families, and our species. From our personal ancestry to the evolution of our species, to our traits, characteristics, and potential disease risks, genomic analyses are providing tremendous insights into what it means to be human.

CHAPTER 8

Genetic Screening

Beyond information regarding ancestry, family heritage, and the vestiges of human evolution, personal genetic testing also holds the power to provide specific insights into potential medical issues. Similar genetic testing and screening that has been normally performed in a medical setting with input from doctors, nurses, or genetic counselors can now be communicated electronically via direct-to-consumer genetic testing. However, as with ancestry information, the relevance and meaning of genetic information for prevention, diagnosis, or treatment of disease can be extremely complicated. Although the information provided by personal genetic testing can be interesting and useful to the individual, personal genetic testing companies might also be quite interested in getting their hands on our DNA for purposes besides simply providing a service to the end user. What exactly is the business model of personal genetic testing companies? In what ways are they monetizing their services?

DIRECT-TO-CONSUMER GENETIC TESTING

Mailing off a bit of spit in a tube doesn't sound like that big a deal. However, even if the sample is not linked to any one person and without any additional information and context, simply having huge numbers of genomes to work with can provide a company with the ability to refine analytical techniques. Large genomic data sets can also help to identify the prevalence of certain SNP patterns among populations and how they correlate. Of course, deeper sequencing of DNA is also possible, and in some cases, are already occurring. If a company can use SNP patterns to provide context and identify ancestry, and then follow on with sequencing specific regions to look for particular gene variants, the value of this combined information both to biomedical researchers looking to identify the genetic basis of human disease and the pharmaceutical companies looking to develop (and profit from) treatments can grow very rapidly. Furthermore, voluntarily providing information regarding your family and personal medical history along with your DNA sample further strengthens the analytical power of your genetic information.

23andMe conducts DNA genetic testing and analysis but also gives consumers a series of voluntary questionnaires that cover everything from personal and family medical history to behavior patterns and personal preferences, like your favorite flavor of ice cream. It would seem that one facet of 23andMe's business model is to assist in understanding the genetic basis to all types of human issues and behaviors. In addition to ancestry and traits such as whether you are an early bird or a night owl, 23andMe provides different levels of analysis regarding disease risk.

Companies like 23andMe aren't working alone. The combination of all this personal information with its accompanying genomic data is extremely valuable to pharmaceutical companies and biomedical researchers, which is why these stakeholders have been playing very active roles in personal genetic testing companies, from publishing research papers together to creating joint initiatives. Some even have direct financial links, such as between 23andMe and GlaxoSmithKline. Clearly, if the outcome of the accumulation of information is of significant benefit to biomedical research and clinical practice, it is hard to argue against the increased prevalence of genetic analyses, particularly if customers are voluntarily participating. Given the popularity of direct-to-consumer genetic tests, it is easy to forget how new they really are. 23andMe has only been testing customer DNA for a little over a decade but already counts more than 10 million people who have used their services. Although these products are becoming more commonplace, we are really only at the start of the new frontier offered by all this genetic information. As with any potentially disruptive technologies and products, it is important to consider the perspectives of the consumer, the companies, and the role of governmental regulatory agencies.

Regulators are weighing in on these services. In 2013 the FDA ordered 23andMe to stop providing health-related information. It took the company three years to gain approval to resume the services by giving evidence that each "test" they were conducting passed muster with regulators. 23andMe providing health information raises important questions regarding the extent to which personal genetic testing should be regulated. Do these kits represent medical devices? What types of risks might be inherent in personal genetic testing? How can providing information to people be bad? What else can be done with all that information?

Privacy is not a binary. It is true that you can opt out of all utilization of your genetic material and information beyond your personal perusal, and most people don't fill out each and every question in every survey provided by 23andMe. However, it's quite natural to want to assist with activities aimed at improving human health, and who really reads all the fine print anyway? But considering how social media companies have exploited personal information in widely publicized scandals should give us some pause. Our genomic DNA is the most personal information there is; it defines who we are on a biological level. Being unique, an individual's genome arguably cannot really be anonymous. On the other hand, our genome is not wholly our own, as it is a combination of our parents' genomes and contains specific signatures that are essentially identical to those of other family members.

Personal genetic testing has been such a boon to people looking for relatives and ancestors because of these overlapping elements. One poignant case is that of Eleni Liff, who was found abandoned in a shopping bag when she was three days old and was finally able to identify her birth parents through 23andMe. However, the shared web of relatives who have had their DNA tested and put into accessible databases also means that genetic information about yourself can be inferred from the genomes of your family members, even distant ones you may have never known existed. These similarities and connections raise some interesting issues.

If your brother finds out that he inherited a gene variant, or SNP pattern, linked to a specific disease, should he be compelled to tell you? What if he chooses to make his genetic analyses public and this leads to you receiving contacts from potential relatives, or marketing materials based upon your potential risk for specific genetic diseases? What if you were unable to obtain certain insurance because a genetic disease runs in your family? It should be noted that while the Health Insurance Portability and Accountability Act (HIPAA) of 1996 doesn't prevent insurance companies from using your genetic info against you, the 2018 Genetic Information Nondiscrimination Act (GINA) does offer significant protections against having genetic testing imposed upon individuals, as well as having any results of genetic testing used against them for health insurance purposes. Protections explicitly created by GINA include preventing restricting enrollment into a health insurance plan based upon the results of genetic analyses and ensuring that health insurance premiums are not based upon genetic information. That being said, other types of insurance, such as life insurance and long-term care, are not covered by GINA. Although these might seem implausible outcomes, without knowledge of the potential uses, and misuses, of this type of information, we might very well find ourselves suffering consequences we didn't intend to sign up for.

How easy is it to interpret the results of personal genetic testing? When it comes to ancestry, the information provided is relatively straightforward: you are x percent such and such nationality. (Of course, understanding the meaning and limitations of ancestry information can be complicated because the analyses and reference databases behind these results are constantly changing, and so can the conclusions.) But the implications for identity, heritage, and family context can be extremely complex, with the potential for life-changing revelations, such as the woman who grew up as an only child only to find out she had twenty-nine siblings (her father was a sperm donor) or the growing numbers of people who are being informed their fathers aren't really their fathers, or that they were adopted. In fact, there are now online support groups where people can share their experiences, such as Facebook groups with thousands of followers who support each other following these types of unanticipated discoveries. There are also complex issues at hand when we consider genes and medical information.

GENES AND MEDICINE

Many medical doctors don't feel adequately trained to interpret genetic results, which is one reason there are genetic counselors who help people understand tests like prenatal screening. 23andMe also recommends speaking to a genetic counselor to get more information and assistance with interpretation of some of the results they provide. Their website can help you find a genetic counselor near where you live.

Broadly speaking, there are two primary types of information going into these kinds of analyses: SNP patterns and sequencing of specific genes. Testing whether someone is a carrier for *CFTR* ΔF508 is one example where individual gene variants can be identified that are causatively linked to a particular disease—in this case, the debilitating and currently incurable lung disease cystic fibrosis. There are many gene variants that have been identified as causative for specific genetic diseases, or at least are associated with increased relative risk, even if the exact mechanism isn't yet understood. Although there are many recessive traits such as cystic fibrosis, which will arise if two mutant copies are inherited, there are also individual gene variants that can't be viewed via this type of simple genetic lens.

Because the metabolic and regulatory pathways in our bodies arise through the interrelated functions of thousands of different proteins, mutation of specific genes can have effects ranging from unnoticeable (silent mutation) to disastrous, such as embryonic lethality (death of the embryo in utero). With a so-called recessive trait, noticeable negative effects can emerge only if two copies of the variant are inherited. Recessive genetic diseases are often the result of a mutation that causes a loss of function of the resultant protein encoded. With one functional copy, health can be maintained within normal limits. In the case of cystic fibrosis, the mutant protein can't properly function at the right place at the right time. However, for people who are "carriers" of only one copy of the mutant variant, enough of the normal copy is expressed so that no disease arises.

In the case of dominant variants, inheriting only one copy of the mutant version of the gene is enough to cause the disease. An example of a disease-causing mutation that displays a dominant inheritance pattern would be Huntington's chorea. This progressively debilitating neurodegenerative disease develops later in life and occurs in individuals who have inherited a single mutant version of the huntingtin gene *HD*. If a parent has the disease, there is a fifty-fifty chance that any child they have will inherit the mutant copy and eventually develop the disease. Thankfully, Huntington's is very rare. But it is an example that highlights the potential benefits of prenatal genetic screening, especially for people with increased family risk, as well as the value of developing genetic tests that can be performed on fetal DNA early in pregnancy.

Many cases of gene variants linked to particular diseases are much more complex than simply being dominant or recessive, and the results of harboring a particular gene variant can also depend on other factors, such as the presence of other mutations, as well as contextual issues. This juxtaposition of genetics and environment is often referred to as nature versus nurture. When it comes to some more complex diseases, especially those that can depend at least in part on environmental factors and individual behavior, such as cardiovascular disease or diabetes, a correlative association with specific SNP patterns can also be analyzed. In one recent report, researchers at Harvard, MIT, and the Broad Institute looked at 6.6 million different SNPs and assessed the risk of several complex diseases, including cancer, diabetes, and heart disease. There are plans to make the analytical framework developed by these researchers freely available to individuals via a web portal that can accept data obtained through personal genetic testing. More targeted analyses of particular SNP patterns that correlate with specific diseases underlie the widespread technique known as genome-wide association studies (GWAS), which will be described further below. Although the general concept of combining huge masses of data to make conclusions about individuals is consistent across most research, the particular types of genetic analyses are potentially relevant to human health and disease, and the depth and utility of this knowledge and understanding can vary tremendously.

When personal data like age, sex, and ethnicity is combined with genetic information such as particular SNP patterns and analyses of specific mutations, 23andMe can tell a lot about an individual. In addition to analyses of ancestry and several types of health reports, 23andMe also provides information about your individual preferences and characteristics, such as whether you are a morning or night person or if you are likely to have a bald spot, or dimples. One nice thing about the way 23andMe describes these characteristics is that in addition to describing the particular types of genetic variants that can play a role in these specific personal details, it provides scientific references that back up its analytical claims. It also explicitly states how it has refined its statistical models through incorporating the responses people have provided to 23andMe's questionnaires, and supplies actual measures of the predictive power of its analyses.

There are two primary types of medical reports that 23andMe generates. The website responsibly requires that you view tutorials describing the general meaning and relevance of each report before you are allowed access to your own results. The 23andMe Carrier Status reports look at a series of genes that have specific variants associated with recessive genetic diseases, such as *CFTR* (cystic fibrosis transmembrane conductance regulator) and identify whether you carry any of these variants. It does not provide a complete analysis of all genes implicated in genetic disease, or all potentially relevant variants for the genes tested. Also, 23andMe states very clearly that it is only analyzing whether an individual is carrying a single copy; it does not specifically test if your second copy of a gene is mutated. This is not in any way meant to be a diagnostic tool. The focus of 23andMe Carrier Status reports are primarily for informational purposes and are of particular relevance for people planning on having children. It is similar to a home-based personal genetic test that shares some aspects in common with a prenatal screen, although with fewer variants tested than many prenatal screens and without the benefit of a genetic counselor to assist with interpreting the results.

Currently, the 23andMe Carrier Status report looks at over forty different diseases and disorders that are generally debilitating recessive genetic diseases. That means that the disease arises only if two mutated copies are inherited and that by adulthood individuals would know if they were suffering from the disease. The information provided is primarily helpful for family planning and understanding hereditary patterns of disease among a group of relatives. If an individual is positive for one of the disease-associated variants tested by 23andMe, or any prenatal genetic screen for that matter, and has a child with another carrier, that means the child will have a 25 percent chance of inheriting the two mutant copies of the gene and having the disease. Similarly, there will be a 25 percent chance the child won't inherit a mutated copy of the gene from either parent and neither be a carrier nor have the disease. Last, the most common outcome, there will be a 50 percent chance that the child will inherit one mutant copy and one normal copy, and therefore be a carrier.

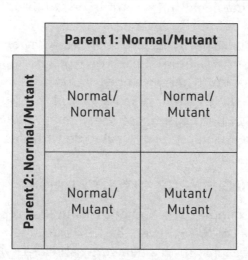

Parent 1: Normal/Mutant

Parent 2: Normal/Mutant		
Normal/Normal	Normal/Mutant
Normal/Mutant	Mutant/Mutant

Box: Offspring of carriers. A Punnett square showing the potential outcomes of two carriers mating, each with a single "normal" and one "mutant" copy of a specific gene. As each gamete, sperm or egg, will only carry one copy of each gene, when fertilization occurs the resultant two copies will have a 25% chance of being both normal, a 50% chance of forming another carrier with one normal and one mutant copy, and a 25% chance of causing two mutant copies to be inherited.

While the Carrier Status reports are primarily focused on "your future family," the Genetic Health Risk report provided by 23andMe focuses on whether you have specific genetic variants associated with increased risk of developing a disease or condition. The Genetic Health Risk report currently tests for increased predisposition to over a dozen different genetic diseases. These risk assessments can be relatively specific, such as an enzyme deficiency that causes a specific genetic disease, for example, the glucose-6-phosphate dehydrogenase (G6PD) enzyme that when mutated can be linked to a specific form of anemia; or they can be more broad and general, such as diseases with a genetic component that can arise through complex pathways and mechanisms, like Alzheimer's or type 2 diabetes. 23andMe makes it explicitly clear that having a specific variant in one of these genes doesn't mean you will definitely develop the disease. Additionally, as not all known variants are analyzed, not having any of the particular gene variants tested doesn't mean you won't develop the disease. In many cases, family history, ancestry, and behavior are relevant considerations, and further discussion with medical professionals is encouraged. Because these tests are not definitive, it does, of course, raise the question of what to do with this information.

WHEN YOU KNOW YOUR POTENTIAL RISK

One specific example in the Genetic Health Risk report generated by 23andMe is the *BRCA1* gene. *BRCA* stands for "breast cancer," and specific variants in this gene are associated with increased risk of developing different types of cancer, not just in the breast. However, some *BRCA1* mutations are very strongly linked with breast cancer. *BRCA1* is referred to as a tumor suppressor gene, and the protein it encodes normally functions to prevent uncontrolled cell division, which can lead to cancer. When there is a mutation that reduces the function of *BRCA1*, cancer can result. Many different *BRCA1* variants exist, and the odds of having a specific variant depends upon genetic ancestry. The specific mutations tested for by 23andMe are more common in people of Ashkenazi Jewish descent. The true relative risk of developing cancer for an individual associated with particular variants can depend on many factors, such as family history and lifestyle. Deciding what to do with this type of information can be very complicated and involve asking complex and difficult questions.

At the extreme end of the spectrum, there are cases where people who have *BRCA1* mutations significantly associated with a high risk of developing breast cancer choose prophylactic bilateral mastectomy—preventive removal of both breasts. This can be a prudent choice for some with particular *BRCA1* mutations, especially if they have a strong family history of breast cancer. However, given the huge number of variants associated with *BRCA1*, and the other genetic, environmental, and behavioral factors that can contribute to individual risk, these decisions can be extremely difficult without significant expert guidance. Just knowing that you have a variant in a gene like *BRCA1* that differs from the general population can be confusing and upsetting, but there are many *BRCA1* gene variants that are not associated with increased odds of developing a disease.

A competitor of 23andMe, MyHeritage, has recently announced its own BRCA test, which differs from 23andMe's in one critical way. MyHeritage employs physicians to review reports. As this means the test isn't truly direct to consumer, the stringency of regulation is somewhat decreased. However, does this really improve the analysis, or does it simply provide a competitive commercial advantage? Another potential competitor to 23andMe, Invitae, recently released data it claimed demonstrated that 90 percent of people with BRCA mutations actually won't be picked up by the tests offered by 23andMe.

What do these tests actually tell us, then? Sometimes a particular variant in a specific gene can be associated with increased risk of a disease, even if it isn't yet clear why. Information directly linking a particular gene variant with a causative role in disease— for example, how a particular amino acid change owing to a mutation alters the function of a protein—isn't always needed, especially for screening purposes, as long as the statistical association is sound. Calculating relative risk between those with and without certain genetic variants does not necessarily state anything definitive about one individual; rather, it looks at comparisons among large populations and synthesizes that information into some kind of likelihood value, or relative risk.

There are many different contexts where sequencing specific genes, or sets of genes, can lead to potentially useful information regarding health. One example would be analyzing DNA from tumors that originate from particular organs that demonstrate well-defined and consistent characteristics—from the way the cancer develops to how it presents clinically, as well as how it responds to particular therapies. Some knowledge about the types of genes that are routinely mutated in such cancers, such as tumor suppressors like BRCA, will allow for specifically sequencing these genes in an attempt to find mutations that could be responsible for that particular tumor. Tumors can harbor mutations not present in the person's genome at birth. These so-called *de novo* somatic variants arise in the body as changes to the genome occur. Often in only a single cell to start with, changes can arise from issues such as chemical or radiation-based mutagenesis and trigger the development of cancer. However, what do you do when you find a variant in a gene, but it has never been associated with that cancer? What about "variants of unknown origin"?

We are full of polymorphisms, and many of them, even those that make changes to a protein sequence, might not have any functional significance. If you don't know how a specific genetic variation might alter the expression or function of the protein encoded by that gene, particularly if it is a new mutation not previously reported, you might simply write it off as a variant of unknown origin. However, what if at some later date further analyses explain how that particular mutation might be involved in a specific disease, or even demonstrate that it is significantly associated with increased risk? Generally speaking, there won't be follow-up with people previously identified as having this mutation before its being defined as relevant to that disease. So while the variant hasn't changed, its interpretation and potential relevance have. Interestingly, this can be one area where the direct-to-consumer personal genetic tests can be potentially more useful over time than those conducted under medical supervision, as companies like 23andMe and AncestryDNA regularly update their analyses and contact customers with new information as it becomes available.

Ideally, genetic testing will be followed by statistical analysis with a large enough sample size to assess for a genetic link, along with detailed biological studies to understand why a specific genotype (genetic background) might be associated with a particular phenotype (measurable or observable characteristics), but it is not always that simple. Providing people with information is well and good; knowledge is power, and all that. If you find out that your genetic background puts you at increased risk for developing lung cancer, you might finally quit smoking. However, if there is nothing you can do about a health risk, what does it matter if you have a slightly higher probability than the general population of developing some complex disease? In particular, if there is no treatment available, how does this knowledge help? Information is not the same as understanding, and it seems as though generating genomic data and analyzing for relative risk values is far outpacing our ability to explain how a particular disease with a genetic component might arise.

One reason for technology being ahead of the curve of our understanding is that historically it has been comparatively much easier to go for the low-hanging fruit of relatively rare genetic diseases caused by mutations that affect the function of one particular protein, so-called monogenic diseases. On the other hand, associating specific SNP patterns with populations exhibiting complex multigenic diseases is also relatively straightforward, and the advent of genome-wide association studies (GWAS) has ushered in the era of big data for medical genetics. However, bridging the apparent gap between these two paradigms is not simple. Overall, there is a spectrum of the magnitude of the effect of a single specific mutation, ranging from extremely subtle to completely catastrophic, combined with the prevalence of that variation that ranges from extremely rare to common. It will usually be easier to identify SNP patterns with more prevalent variants that have greater functional impact than those that are less common and have relatively minor individual effects. These variables will translate into differences in the number of individuals required in any study, as well as the number of SNPs that need to be assessed to generate useful data. Some recent evidence suggests that rare genetic variants, including those that could cause disease, can readily be picked up as false positives on SNP arrays, including some of those used by personal genetic testing companies. This raises concerns about the level of confidence in the results presented to customers, particularly if the methods for confirmation of potential positive findings are not readily apparent.

Even if the results of a SNP array are confirmed, understanding how different variants of specific genes can alter the relative risk of developing a complex disease modulated by other issues such as ancestry, family history, and behavior is extremely complicated. Assessing all the variables is proving challenging for medical doctors and genetic counselors, so how can individuals who purchase personal genetic testing kits hope to truly understand all the information provided, as well as its limitations? As whole genome sequencing becomes more of a reality in a medical context, are we prepared to attempt to understand the potential relevance of all differences between an individual and a "reference genome," and how can that be defined given the complex ancestry of many individuals? Considering how little we truly know about the function of many of the thousands of proteins encoded in our genomes, a complete and full link between genetic analyses and biomedical understanding is a long way off.

A holistic understanding of the relative costs and benefits of conventional genetic testing in a medical context relative to personal genetic testing kits would be extremely complex and involve many factors. As described above, some personal genetic testing companies may well be enticing customers with low costs to capitalize on the information provided in lucrative collaborations with separate entities like pharmaceutical companies. Although alternative uses of an individual customer's genetic information generally won't be performed without consent, the health insurance and medical diagnostic industries, not to mention health providers, are driven by a profit motive. That being said, consulting with experts can certainly assist with understanding complex information, and for many people the results of a genetic test are well outside their general knowledge base. However, not all medical professionals are experts in genetics, and bringing the results of a personal genetic test to your doctor can potentially lead to frustration on both sides.

The combined power of detailed genetic information of specific individuals and huge reference databases is being explored by many researchers, medical doctors, and pharmaceutical companies. From personalized medicine, where your individual genetic background guides medical decisions, to gene therapy, where dysfunctional genes can potentially be corrected through manipulations on a genetic level, and pharmacogenomics, where particular drugs are selected or avoided based on information present in the patient's genome, the power of genetic information is poised to revolutionize health care.

Can data generated from personal ancestry and family history genetic testing be applied to questions in other disciplines? Beyond medicine, an interesting application of the availability of human genetic information in publicly accessible databases is assisting investigators with solving cold cases.

CHAPTER 9

Genomic Justice

G enomic information has been employed in criminal investigations in various forms for many years. Nearly twenty years
ago, a searchable database accessible to police called the National
DNA Index System (NDIS) was created as part of the FBI's Combined DNA Index System (CODIS). NDIS currently holds over ten
million profiles of people arrested or convicted of crimes, including
DNA data. Although this resource is national, the implementation
of NDIS is regulated at the state level. Nevada, for example, passed
a law in 2013 requiring the collection of DNA from all felons.
Although certainly an extremely useful tool for law enforcement—
in light of the prevalence of recidivism—there's something a bit
Orwellian about compelling citizens, even convicted criminals, to
provide DNA samples for collection and databasing. That being
said, through a wide variety of mechanisms of collection, analysis,
and utilization, NDIS has been instrumental in solving crimes and
catching many dangerous criminals. It should not be forgotten that
DNA testing has also exonerated the falsely imprisoned.

Recently, the explosive growth of personal genetic testing
has opened up entirely new avenues for using genetic information in criminal investigations. This treasure trove of potentially
useful information has raised some new concerns over personal
privacy and the potential for investigators to wade into murky
ethical territory. The Genetic Information Nondiscrimination Act
(GINA) and the Health Insurance Portability and Accountability
Act (HIPAA) provide some protections against some potential

concerns (see p. 108). But only now is the admissibility of evidence gained from personal genetic testing being assessed in court. The types of regulations that might be implemented for the collection, analysis, and use of personal genetic testing data for criminal investigations remain an open question.

An interesting perspective on the use of genetic genealogy databases for investigative purposes is that the vast majority of people performing personal genetic testing until now have been of white European descent. As searches for DNA matches are objective and blind to racial prejudice, some of the discrimination prevalent at every level of the criminal justice system might be countered through further application of these tools.

Private companies such as AncestryDNA and 23andMe have reserved the right to restrict access to their databases for use by criminal investigators unless compelled by court order. Recently, FamilyTreeDNA actually issued an apology to its users for assisting the FBI with cold cases without disclosing that fact. GEDmatch, on the other hand, has recently been in the news a great deal as they have opted to permit investigators to use the information they maintain. The company's terms of service expressly allow its users' raw data uploads to be compared with "DNA obtained and authorized by law enforcement to identify a perpetrator of a violent crime against another individual . . . [or] to identify remains of a deceased individual." GEDmatch users can opt out of these types of comparisons, removing their genomic information from the GEDmatch database, or deleting their accounts altogether, at any time.

The first thing to understand is that GEDmatch is not a personal genetic testing company; it doesn't perform any laboratory work. Rather, it is a depository for the information provided for users of personal genetic testing companies that individuals can utilize to analyze and understand their results. GEDmatch is an online resource developed for genetic genealogists, both amateur and professional, that is broadly open access for uploading and searching. Before the advent of personal genetic testing, people interested in learning more about their family trees or looking for potential relatives made use of a variety of sources such as birth and death records and census and immigration data. These could be used by those seeking simply to understand personal family history as well as by people who were adopted and want to find their biological parents or long-lost siblings.

GEDmatch emerged organically from the community of genetic genealogists and exists as a searchable database with multiple analytical resources where individuals can upload and cross-reference their genetic information regardless of the source. GEDmatch is free to join and use, although access to some advanced features is only available to people who pay a fee. Information from virtually any personal genetic testing company can be uploaded to GEDmatch and used to investigate ancestry and search for potential relatives. However, armed with the ability to search among the millions of individuals with information available in the GEDmatch databases, genetic genealogists assisting criminal investigators have begun to have amazing results solving cold cases.

DNA matching to identify criminals, or exclude potential suspects, has been used in law enforcement for quite some time. If there was a likely suspect and there was a source of DNA at a crime scene potentially belonging to the perpetrator, all that was needed is to get DNA from the "person of interest" and see if the two samples match. This is quite straightforward if you know from whom to get the DNA. But what if there are no suspects? That is where GEDmatch has come into play and been extremely useful.

Keep in mind that your DNA is not wholly your own. You share it with your relatives. On average, your genome will be 50 percent identical to each of your parents, and also your siblings. You are 25 percent identical to each of your grandparents, as your parents are 50 percent related to their parents, and so on. You are also 25 percent identical to your aunts and uncles, and thus 12.5 percent identical to your first cousins, as they are 50 percent related to their parents, your aunts and uncles. As you go further beyond your nuclear family, the percentages might get lower, but these levels of relatedness are still very significant for investigative purposes. By combining these types of analyses with traditional genealogical techniques, crime scene DNA data uploaded into GEDmatch can very rapidly be used to identify potential suspects . . . if one knows how to look. Genetic genealogists have the skills and the techniques to make the links.

Once the genetic information obtained from the crime scene corresponding to the unknown criminal is uploaded to GEDmatch, relatives within the database can be identified. The genetic genealogist will then attempt to fill in the rest of the family tree using all the tools at their disposal. This often includes standard genealogy approaches, such as searching birth, marriage, and death records, in combination with the genetic information. The ultimate goal will be to identify a person who could have that degree of relatedness to those matches to individuals in the database, and was alive and located in the general vicinity where the crime took place. If someone fitting these criteria is identified, police can interview the potential suspect. They can then attempt to determine if there is a match between that person's DNA and the crime scene sample.

COLD CASES . . . CLOSED

These procedures have been successfully employed with several high-profile cold cases. The Golden State Killer was suspected of a gruesome series of rapes and murders over an approximately twenty-year period ending in the mid-1980s. Finally, in 2018, with significant assistance from a genetic genealogist, a suspect was arrested and charged with thirteen counts of first-degree murder. In another case, genetic genealogy has recently led to the conviction of a man for a double murder that took place over thirty years ago. Interestingly, at trial the murderer's lawyer did not challenge the use of these techniques on the grounds of privacy and informed consent. In fact, at least one DNA testing company is now using the potential to assist in solving cold cases as a marketing tool. This relative acceptance of the techniques could be because although they might be revolutionary to police and the greater public, from the perspective of the genetic genealogist, assisting in cold cases seems to be a natural extension of the tools and techniques they have already been employing for other purposes, such as finding long-lost relatives.

Although it is hard to argue against using existing information to catch criminals and solve cold cases, pursuing suspects in this way does raise some questions of ethics and privacy. Where do we draw the line? If it is OK to do this for murder, what about assault? What about spitting on the sidewalk or stealing a cup of coffee? If you are wrongly contacted by the police because some distant relative you may never have met committed a crime, is that inappropriate or an undue burden? According to a recent study, albeit small in scale, approximately 60 percent of people of white European descent can expect to match with an individual in a genetic database who would be a third cousin or closer relative. Just because a few distant relatives chose to upload their genetic information onto GEDmatch, does that mean you consented to being involved as well? What about the means by which investigators get a DNA sample from a person of interest? In some cases, this involves swabbing a cup or napkin discarded in a trash can. These same methods have been applied for years to identify criminals when a suspect is known; the only difference now is the investigative tools leading to that determination—in other words the person of interest is identified through a family DNA match and was unknown to investigators before that connection was made. Increasing the assessment of individuals identified through these nontraditional means does raise questions. However, it is hard to argue against the results (apprehending criminals who would otherwise have gone free) and it will be extremely interesting to see if the approaches of genetic genealogy become standard parts of the criminal investigator's arsenal.

Although large forensics laboratories with extensive technical expertise perform the bulk of crime scene analyses, some local police forces are taking matters into their own hands and purchasing and operating instrumentation for generating DNA profiles, which raises questions about the use, and the potential misuse, of these sophisticated instruments. Although the operation is relatively simple—about as challenging as running your dishwasher—these systems are generally meant to be employed for cheek swabs and other similarly simple and relatively pure samples. Basically, these instruments can extract DNA from a sample and then perform PCR with primers that amplify certain regions of genomic DNA termed "short tandem repeats" (STRs) that vary in length among individuals. This analysis can generate a unique DNA fingerprint in less than two hours.

These instruments might not have been specifically designed to analyze complex sources of DNA, and sometimes they may be used for crime scene evidence, which can be significantly less homogenous and concentrated than conventional sources of DNA like saliva samples. Furthermore, sample contamination and technical issues with the instruments can hamper data collection. If the system cannot generate a usable data set from a particular source of DNA, that sample is then no longer available for alternative techniques that might have otherwise been successful.

Although some STR-based DNA profiles generated within police departments can be uploaded to CODIS, many are being added to more local registries. Although police must request permission before they can take a cheek swab from an individual, few people seem to refuse. Generally, data sets generated from individuals are compared to DNA profiles from crime scene evidence in attempts to identify and exclude potential subjects. This is still more of an investigative tool, rather than evidence that might be presented in court. Imagine a scenario where the local police have a DNA profile for everyone in town in their database just waiting to be matched to a crime scene or some piece of evidence. It does seem a bit dystopian.

Apart from investigations into criminal activity, our unique genetic profiles—for example, from SNP arrays—have the power to tell us about our ancestry, our health, personal preferences, personality and behavioral traits, and many other types of interesting and important information. However, what about a deeper understanding about the root causes of disease? What can genomic analyses tell us about the specific mechanisms responsible for genetic diseases? How were the gene variants assessed in prenatal screening or the 23andMe Carrier Status reports originally identified? What can the search for these rare mutations tell us about normal biology?

CHAPTER 10

Classical Methods to Search for Disease-Causing Mutations

There are a few different techniques employed by biomedical researchers looking for the mutations that cause disease. Completion of the Human Genome Project in 2003 supported huge advances in these endeavors by generating an overall understanding of the genome that became a reference map with standard sequences for all potential disease-causing genes. However, even before completion of the human reference genome, or the ability to generate whole genome sequencing, or even large-scale sequences of only the expressed genes (the exome), innovative scientists developed creative means for identifying disease-causing gene mutations.

Born almost two hundred years ago, Gregor Mendel is often referred to as the father of modern genetics. He performed extensive plant breeding studies at a monastery in the current-day Czech Republic and determined several fundamental laws of genetics. One of the observations that the monk Gregor Mendel made through his studies of pea plants was that different traits were not necessarily inherited together in any predictable fashion. When Mendel had pea plants with different characteristics such as long or short stems, or purple or white flowers, each trait would be inherited independently of the other. There was no common causative link connecting attributes like stem size and flower color. Similarly, although you and your brother both have red hair like your father's, it doesn't mean you will both also definitely have his long nose.

This phenomenon is called the law of independent assortment, which basically states that different genetic traits are inherited independently of each other. It should be noted that this is usually employed to describe differences in traits caused by different variants of single genes, commonly referred to as simple Mendelian inheritance. So Mendelian inheritance is the basis for understanding the dominant and recessive effects previously described above, and the law of independent assortment states that the traits inherited through these genetic means can broadly speaking be studied one at a time. Each trait that depends upon one or another dominant or recessive genetic variant is independent of all others. As most recognizable human traits are actually caused by the combined activity of multiple gene products, analogies like those described above are for illustrative purposes only. However, monogenic (single gene) traits that exhibit Mendelian inheritance are very important for understanding human disease, at least those caused by single gene mutations.

Imagine that two people who are both carriers for two different recessive diseases have a baby. Let's call A and B the non-mutant, nondisease-causing variants, and A* and B* the mutant and disease-causing variants. As carriers, both parents would be A/A* and B/B*. The law of independent assortment states that the risks of the baby inheriting two mutants of either gene (A*/A* or B*/B*) are completely independent of each other.

This is because there are two copies of each gene in the genome, one provided by the mother and the other from the father, and they can have different sequences. These differences can cause alternative forms of a trait. Generally, one will be dominant and the other recessive. In the example above, A* and B* are recessive, and two copies of either are needed for the baby to have one of the diseases. A and B are dominant, and that is why the parents' genotypes, A/A* and B/B*, make them carriers who do not display the recessive disease trait.

One of the things that Mendel noticed was that when individuals with different versions of a monogenic trait are crossed through sexual reproduction, either one or the other form of the trait will emerge in each individual offspring. Usually, in the case of monogenic traits, intermediate forms of the characteristic are not observed—for example, breeding together a tall plant and a short plant results in offspring that are either tall or short, not plants that display intermediate height. So, height is the monogenic trait and short or tall are the variables, and one of the variants will typically be dominant over the other. If you inherit one copy of each version of the gene, you are said to be heterozygous, and will display whichever is the dominant trait. Only if you inherit both copies of the recessive variant will you display the recessive trait. Of course, if you inherit two copies of the dominant variant, you will also display that trait. The pea plant example is admittedly highly simplified, and human height is a spectrum. Although certain single gene variants can be associated with extreme deviation from the average, many subtle genetic and environmental variables have an impact on the height of an individual human.

DOMINANT OR RECESSIVE?

Recessive traits aren't always associated with a disease. They are often simply an alternative expression of a variable that may not even have any significant contribution to overall fitness, at least not for humans in the modern world. Imagine there is a gene that can be in two forms, A or a, and that A is the dominant version, and a is the recessive (a is used for the recessive, as opposed to A^*, to show that the recessive variant is not disease causing in this example). Thus, if two heterozygous (A/a) individuals are crossed, 50 percent of the progeny will also be heterozygous (A/a), 25 percent will be homozygous dominant (A/A), and 25 percent will be homozygous recessive (a/a). Thus, the recessive trait will only be displayed in the one-quarter of individuals with a/a, and all other combinations will display the dominant version of the trait.

	A	a
A	A/A	A/a
a	A/a	a/a

In pea plants, tall stems are dominant, and short stems are recessive. That means that if a pea plant inherits one copy of the variant associated with tall stems and one copy of the variant associated with short stems, the pea plant will be heterozygous and display the dominant trait, tall stems. Only a pea plant that inherits two copies of the variant associated with short stems will display the recessive trait, short stems.

In the case of mammals like us, sperm and eggs are formed during the process of meiosis (cell division), and each holds half of the total DNA required to make a cell. When fertilization occurs, a complete genome is formed again. Humans have twenty-three different chromosome pairs, a total of forty-six chromosomes, and we inherit one chromosome from each pair from our mothers and the other from our fathers. So if two genes are on two different chromosomes, say one on chromosome 5 and the other on chromosome 9, whether a sperm or egg receives the maternal or paternal copy of those two chromosomes will be completely independent of each other. Thus, traits are independently assorted.

However, what happens if the genes are on the same chromosome?

There is a phenomenon referred to as recombination that occurs during meiosis and greatly increases genetic diversity. In summary, the maternal and paternal pairs of each chromosome line up next to each other like two trains stopped across from each other at a subway station and exchange genetic information. This occurs because although the specific variants on each chromosome might be subtly different, the same genes are arranged in the same order, so genetic information can be transferred from one to the other without any issues. Just because one of your chromosomes had two specific gene variants on it doesn't mean that same chromosome will have those two specific variants once it makes its way through meiosis.

If you have one chromosome with the A variant of one gene and the B* variant of a second gene, and if your other paired chromosome had the A* variant of the first gene and the B variant of the second, at the end of meiosis the chromosome that actually ends up in a sperm or egg might have any combination of A or A* for the first gene, and B or B* for the second (A/B, A*/B, A/B*, or A*/B*). This unpredictability is what is behind the law of independent assortment. However, there is an important potential exception to this law, and it has proven extremely relevant to the search for genes that cause rare human genetic diseases.

RARE GENETIC DISEASES

Recombination occurs only if the maternal and paternal copies of two chromosomes are physically close and aligned so that the same genes are arranged side by side. However, the machinery of recombination doesn't simply swap the sequences specifically corresponding to individual single genes. Any other DNA that is immediately adjacent will also be transferred between the two chromosomes. Clever genetic researchers have made use of this in their efforts to identify specific recessive mutations that cause monogenic diseases by employing a technique referred to as positional cloning.

Positional cloning has been performed for over thirty years and numerous recessive mutations that cause specific monogenic diseases including cystic fibrosis, polycystic kidney disease, Huntington's disease, and certain types of muscular dystrophy have been discovered through this technique. The basic procedure for positional cloning involves studying DNA samples from people with the disease, ideally from within a large family group where some people have the disease and others do not. There are certain known sequences within the human genome referred to as genetic markers. These can be SNPs or certain types of repeating DNA sequences such as those referred to as "microsatellites" that are located at specific sites in the genome and vary within the population. The different variants of the markers generally have no functional consequences; they are merely used to locate the region in the genome where the disease-causing mutation might be found. If a large number of markers (thousands) are examined, the location of each in the genome is known, and they vary within the population, it's possible to find the mutation.

The basic idea is that DNA is taken from people with and without the disease and assessed to identify which markers are always found in the people with the disease, but generally are not found in the people who do not have the disease. Analyzing large family groups of multiple people with the disease ensures you are looking for the mutations in the same gene. Studying DNA samples across multiple unrelated individuals might lead to combining separate diseases with similar symptoms caused by mutations in different genes. Furthermore, as these gene variants are thankfully quite rare, these monogenic recessive diseases are often more common in families that are particularly inbred—for example, those in which cousin marriages are common.

Although there can be different specific outcomes of this type of investigation that may require some complex statistical analysis—such as the combined action of mutations in more than one gene—the most straightforward result is a single marker that is always found in people with the disease. The genomic location of the mutated gene causing the disease should be very close to the site of a single marker always associated with the disease, as that is the only way that the marker and the mutation will always be inherited together, and it is this proximity that allows the law of independent assortment to be broken.

Once a marker tells you the general genomic location of the mutated gene has been found, the next step is to specifically identify the genetic cause of the disease. Before the completion of the human genome, this could mean a great deal of painstaking hard work. If no genes that might play a role in the specific disease process were known to exist near that marker position, the only possible approach was to start sequencing from that point out in both directions and look for genes that showed polymorphic differences in the people with the disease, relative to people who didn't have the disease, and also seemed to have potentially functional relevance to the disease process. Historically, this was performed by standard Sanger sequencing and the process was painstaking, long, and expensive. It should be noted, however, that by combining samples from multiple families with the same disease, different mutations in the same gene that arose completely independently can be identified. This can increase the complexity of the analysis to some extent, but it also increases confidence that the gene responsible for the disease has been found.

It is possible that there could also be polymorphic gene variants not directly responsible for the disease that happen to be near the marker identified, so in order to confirm that the causative mutation has been found, a functional experiment would generally need to be done. An assay could be conducted in cultured cells showing that the mutant version of the protein that was identified demonstrated functional differences to the normal nonmutated version, or demonstrating that the gene is definitely expressed in the tissue where the disease originated. Understanding the biology that leads from a specific mutation to a complex human disease can be important for verification that the search identified the correct variant, as well as potentially helping in the search for treatments and cures. Confirming the particular gene variant that is responsible for the disease could also mean creating an animal model expressing the mutant protein. One way this is performed is by introducing the potential disease-causing mutation into the genome of a mouse line and then performing detailed biomedical studies of the mutant mice. Animal models exist for many human diseases; some have been engineered, while others have been identified by screening for disease-associated characteristics—for example, a particular line of mice or rats that displays abnormally high blood pressure or cholesterol levels. If an animal-based experiment resulted in a disease similar to what occurred in the human patients, that would be excellent confirmation of the particular disease-causing gene.

Now that the human genome has been completed and we basically know the location and sequence of every gene, the identification stage of positional cloning has been greatly simplified. All that is needed is to search on the human genome database which genes are known to be near the site of the marker, and then it will be possible to see if any might be responsible. However, in general, experiments to confirm that the mutation is definitely causative for the disease still should be performed.

Many mutations underlying monogenic recessive diseases have been uncovered through positional cloning. Furthermore, through analysis of multiple diseases with similar symptoms entire pathways underlying critically important physiological processes have been uncovered. One area where this has been incredibly informative has been increasing the understanding of how salt and water are handled by the kidneys, and how diseases such as high blood pressure can result when genes in these pathways are mutated. Not only can positional cloning tell us about individual diseases caused by mutations in specific genes, but when combined together, multiple related analyses can inform us about the overall pathways controlling important processes. Once it is known how several genes that work together in a pathway can be significantly inhibited by specific rare mutations, it is possible to use this information to begin to understand how subtler genetic changes might be responsible for normal variance of a trait within a population beyond simple monogenic recessive disease.

A STUDY OF BLOOD PRESSURE

Richard Lifton, a biochemist who is now the president of The Rockefeller University in New York, used identification of specific rare mutations, largely through positional cloning, to uncover a great deal about how blood pressure is regulated. Lifton studied a large number of familial diseases that cause severely altered blood pressure and discovered a series of mutations in genes responsible for salt excretion through urine. The general idea is that if you can't excrete salt in your urine, it will stay in your blood, and so will more water to maintain balance, but this results in an increase in your blood volume and blood pressure. Lifton also studied people with very low blood pressure, who excreted too much salt, along with water, in their urine.

This discovery allowed Lifton to create a catalog of genes that when mutated in ways that greatly reduced function would lead to potentially deadly and debilitating blood pressure issues. Thankfully, these individual mutations responsible for potentially deadly monogenic recessive diseases are very rare. However, high blood pressure (hypertension) is one of the most common medical problems in the world. For the vast majority of people with hypertension, the disease won't be as drastic as for the individuals Lifton studied. Generally speaking, in most people with high blood pressure, there are multiple genes, as well as behavior and environment, that contribute to the medical problem. By combining the knowledge gained from multiple separate studies of different genes that can cause catastrophic blood pressure issues when mutated in specific ways, more nuanced analyses of less significant variants can be conducted. Putting together these types of observations, a great deal of information has been gained regarding the different pathways through which blood pressure is regulated. Polymorphic variants of the same genes that might not individually be enough to cause significantly debilitating disease can be screened, and from this type of pathway analysis those at higher risk for developing high blood pressure can be identified, and potentially treated to address the possible causes of their specific issues.

Positional cloning generally makes use of a three-step procedure to arrive at the functional basis behind a specific recessive monogenic disease. In the first step markers that are known specific sequences found at particular places in the genome of some, but not all, people are used to identify the general chromosomal location associated with the mutation site. This is based upon the exception to the law of independent assortment described above, where two genomic DNA sequences very close together will be inherited together following recombination. In the second step, DNA sequencing is performed to determine which gene variant might be causing the disease. In the final step, experiments test the hypothesis and hopefully provide functional validation. These experiments can be conducted in cell culture or animal models—for example, through generating a transgenic mouse line containing the mutation being studied. Knowing the sequence of the human genome has greatly facilitated this process, particularly DNA sequencing.

Other genomic techniques have also been applied in the search for mutations behind human genetic diseases. One alternative method that has become significantly more straightforward since the completion of the Human Genome Project involves analysis of so-called candidate genes that could be behind a specific disease. We have a good deal of information regarding the functions of many genes and can infer the function of others from their similarity to well-understood proteins, or even simply to the tissues and cell types in which they might be expressed. There are countless examples where researchers have identified polymorphisms in genes that seem good candidates for specific disease processes. Unfortunately, many of these candidate gene studies have not been successfully reproduced, and further analyses have demonstrated that the variants identified were not actually causative for the disease that was being studied. One area where this has been particularly apparent is in the search for the genetic basis of psychiatric conditions. As these types of issues can be difficult to quantitatively measure, this potentially suggests that the ability to specifically and consistently diagnose a disease is critical to the reliability of the ultimate analysis.

Although positional cloning, when performed correctly, can specifically link genetic variations to particular functional consequences underlying a disease, it is generally limited to rare monogenic recessive traits. Not all diseases are monogenic. Cancer, heart disease, and diabetes are examples of common complex polygenic diseases that can be caused by contributions from multiple different polymorphisms in combination with environmental and behavioral factors. What if you could uncouple the search for the genetic basis for disease from the actual functional causes, removing the mechanistic biomedical portion of the analysis, and simply focus on the medical diagnosis and the genetic association? What if instead of identifying mutations that alter protein function and cause disease, you focus on identifying a genomic signature that could be used to screen for a disease? Simply by looking at multiple SNP markers in large populations, insights into complex polygenic diseases can be obtained, and this is what is behind the technique generally referred to as genome-wide association studies (GWAS).

CHAPTER 11

Genome-Wide Association Studies

There are a large number of distinct genomic tools currently being developed and employed for use in disease screening, prevention, diagnosis, and treatment. Understanding the different roles of genetic information in medicine requires some philosophical considerations. How much emphasis should be placed on understanding the root mechanisms of a disease, compared with the causes of specific symptoms, and finally selecting appropriate treatment?

If you have a persistent cough because of a bacterial infection, and a particular antibiotic will make it go away, does the doctor need to know precise molecular details regarding exactly which bacterial species has colonized your lungs? Is it important to know how and why your immune system wasn't able to clear the infection on its own? Does it really matter at all if the doctor can explain exactly how a bacterial infection makes you cough, or even how the antibiotic will function to kill the bacteria? As long as the antibiotic works and clears up your infection, you are all good, right?

Although it might be critical in developing diagnostic tests or potential therapeutic interventions, knowing the precise proteins on the surface of the bacteria that are stimulating your immune system to fill your lungs with mucus doesn't really matter much in a standard clinical setting. The primary goal for most clinicians is to identify the appropriate medical response to successfully treat the patient, without causing harm. Doing so requires knowing if the bacteria in question is susceptible to a particular antibiotic, and that the patient is not allergic to that antibiotic. Understanding the causal mechanisms underlying disease is not as important to the clinician as selecting the correct intervention.

Adopting a broader perspective has also been done when assessing the genetic basis of disease, or other complex traits or conditions, by using the technique known as genome-wide association studies (GWAS). At the most basic level, GWAS involves taking samples from very large numbers of people with and without specific diseases or conditions and performing analyses of the different patterns of single nucleotide polymorphisms (SNPs)—single base changes at specific places in the genome—among the groups compared. Approximately ten million of the six billion base pairs in the human genome display polymorphism—that is, can exist as at least two different nucleotides. In many cases, there is no currently detectable functional difference between the potential specific polymorphic sequences. The general idea is that by looking at large numbers of SNPs, patterns can emerge that can be useful for clinicians. GWAS is a statistical association or correlation, and therefore does not necessarily provide any information about causation. That being said, GWAS certainly can provide information assisting with biomedical investigations of the molecular basis for a disease, but this is not an essential component of the overall approach.

One major goal of GWAS is to develop tools for evaluating individuals who might be at risk for a particular complex disease with a genetic component. This could be done for both diagnostic and screening purposes. Sometimes different diseases that respond to distinct treatments have similar symptoms. With the ability to differentiate a genetic signature between two similarly presenting diseases, techniques like GWAS could assist with determining which disease a patient might have, and therefore what the appropriate treatment might be. Alternatively, for people who have yet to develop the symptoms of a disease with a genetic component, GWAS could be used to predict whether they might have increased relative risk compared to the general population. As GWAS makes use of patterns of large numbers of SNPs, this approach can be much more powerful than the direct assessment of different variants of a single specific gene. However, GWAS does not necessarily provide specific information about particular disease mechanisms.

GWAS has been actively pursued by researchers for over a decade, in many cases with thousands of individuals analyzed per study. Generally speaking, fewer individuals are required in a GWAS study if the variants tested have more individual causal impact or if they are more common within a population. This is because the definition of a mutation versus a polymorphism is largely semantic. Although GWAS doesn't focus on causation—how any specific variant alters expression or function of a gene—the polymorphisms potentially relevant to a disease should have some direct effect, and the greater this is, the easier it should be to find them. As with most genomic analyses, genetic homogeneity within the study group tends to reduce the inherent variability (and thus number) of individuals required to obtain significant results, but at the same time will decrease the overall applicability of the observations across groups with differing ancestry.

Although GWAS does not directly focus on identifying the root cause of a disease, it can lead to hypotheses relevant to the search for causal mutations that underlie the origin and development of disease. That being said, there is a wide gulf between raw GWAS data and direct understanding of how diseases originate. In particular, many of the SNPs that are assessed are not actually found within the coding regions of genes that determine protein function. Of course, regions of the genome adjacent to gene sequences can have functional significance in regulating how, where, and when a gene is expressed. Polymorphisms in so-called enhancer and repressor elements—genomic DNA sequences involved in increasing or decreasing expression of specific genes, respectively—can certainly have an impact on disease processes.

The technology behind GWAS has become extremely advanced since these studies have begun. Many researchers, large research consortia, and biotechnology companies have developed tools and technologies for performing GWAS and have applied them to huge numbers of individuals with the goal of better clinical outcomes for people at risk for a variety of diseases, including heart disease, cancer, and type 2 diabetes. SNP arrays—small analytical tools that contain many short DNA sequences that match multiple SNPs—are one type of technological innovation often employed for GWAS. These can easily and quickly be combined with human DNA samples and then read automatically, providing easy-to-analyze quantitative data. Disease-specific SNP arrays that particularly test for subsets of SNPs associated with individual complex diseases, such as cancer, have been developed. In this way, out of the general broader cancer pool, a particular type of cancer can be analyzed for a specific SNP pattern.

The lack of direct information about the mechanism of a disease is not the only limitation of GWAS. If only a few SNPs are significantly specifically associated with a particular disease state, it can take a large number of samples to determine a definitive SNP pattern with complete statistical certainty. Also, there can be limitations to what can be determined if the particular GWAS is not being performed for a binary, black-and-white determination between those with and without a disease but, instead, for a more nuanced or graded characteristic. There is a big difference between studying two fixed groups, one with clinical diagnosis of high blood pressure (above a certain threshold) and the other with normal blood pressure, and studying anyone and everyone along the entire spectrum of blood pressure values.

GWAS is being used for multiple human diseases, conditions, and traits. It is relatively simple to imagine comparing two populations, one with a specific defined disease diagnosis, and one without. However, GWAS is also being employed to analyze the genetic bases behind many different complex human traits, such as height and weight. If dealing with two broadly distinct populations—say, adults under five feet and adults over six feet—it might be quite straightforward to perform GWAS to identify SNPs with a very large relative association with height. However, if working with a group of people in the attempt to break it down into groups varying by, say, an inch in height, it might be impossible, statistically speaking, to collect enough samples from each group to make any definitive conclusions about the impact of specific SNP patterns on height.

GWAS can be strengthened by incorporating additional information about the individuals analyzed—for example, clustering familial groups within data sets. That being said, families also often share environmental conditions, as well as genetics. Therefore, this method can pose problems. If you were to simply perform GWAS on two different families (one with a disease and one without), your results would likely include a number of SNPs that vary between the two families but have nothing specifically to do with the disease in question.

GWAS is certainly a robust and efficient means to obtain clinically relevant information regarding complex diseases with a genetic component, in addition to other traits that vary according to expression and function of specific genes. However, GWAS has significant limitations that prevent direct understanding of the disease processes through simple SNP-pattern associations. Employing individual GWAS data sets to calculate one specific individual's risk of developing a complex polygenic disease can be extremely challenging. Although relatively straightforward and useful as a research tool, application of GWAS to clinical screening and diagnosis is proving very complex.

On the whole, the results of GWAS are highly reproducible, which is somewhat of an exception in studies linking genomic signatures to disease risk. In particular, many studies focused on identifying variants of specific candidate genes that have been selected for their presence in a pathway potentially linked to a specific disease state have not been as reliable as GWAS. But one question still arises with GWAS: At the end of the day, how useful is risk data beyond informing doctors and patients? While it can be helpful as a screening tool to identify those at risk for developing a specific disease, particular SNP patterns identified through GWAS don't naturally lead to further direct therapeutic intervention.

Even when we focus on SNPs within protein-coding regions, which are usually a minority of the total, the contribution of any individual gene polymorphism to a particular trait or risk of developing a specific disease will be extremely small, as many of the most prevalent diseases (e.g., heart disease and stroke) are not caused by a single strong mutation but many small differences in the context of particular behaviors and environments. The power of GWAS is the combination of the number of SNPs assessed and the population size that can be analyzed. Determining alterations in protein structure and function derived from the SNPs identified in GWAS is generally not a realistic goal. If a SNP is outside a protein-coding region and instead modulates the expression levels of particular genes, this can be even harder to decipher, especially if the functional significance of a potentially small change in expression is unknown. Although it may be relatively straightforward to figure out if a specific SNP is within a gene, or possibly an enhancer or repressor element, the sheer number of potentially relevant SNPs identified in a single GWAS generally makes deducing the overall impact from this type of analysis unfeasible. Furthermore, our functional knowledge of much of the genome is extremely low, so quite often we won't have a great deal of insight into the meaning of any particular SNP. Without specific actionable findings such as particular gene variants that can be targeted with specific therapeutics, what are the possible next steps after GWAS identification of SNP patterns associated with disease risk?

A general tenet of the scientific method is that replication is required for observations to be broadly accepted and applicable to further investigation. When it comes to GWAS, experimental replication can consist of similar studies performed on different populations selected according to statuses like age, disease progression, or ethnic background. Particularly strong associations can be maintained across different study populations, and this is very good evidence of connections between those SNPs and overall disease risk. Large-scale metastudies that combine various GWAS—such as the analyses of different study populations for risk of the same disease, or a study that combines different diseases that might share common genetic causes (e.g., autoimmune diseases)—can provide further statistical analyses and identify relevant and reproducible findings. While lower levels of irreproducibility in GWAS relative to candidate gene approaches translate into fewer false positives, the corollary is that the low impact of many individual SNPs on disease risk means that GWAS most likely will have a high degree of missed variants that would be significantly associated with disease risk if larger numbers of individuals were studied.

At the other end of the spectrum from GWAS in terms of efficiency and simplicity is the technique of whole genome sequencing (WGS). To take a critical look at WGS, it's important to pose a number of questions: WGS can provide vast amounts of potentially relevant data, but how much of this is definitively useful? How much does the utility of WGS depend upon our current understanding of gene expression and protein structure and function? Does the most efficient genomic tool for clinical applications depend on the disease in question? What if the goal is disease screening, prevention, diagnosis, or treatment? How does the search for answers change as we learn more about the structure and function of specific proteins and the normal cellular function of the complicated interdependent machinery underlying human health and disease? As the efficiency of genome sequencing in terms of cost per nucleotide improves over time, what is the impact on what these studies will reveal?

As will be described in the next chapter, various analytical and therapeutic approaches are being developed with distinct genomic tools, each with particular strengths and limitations. It is important to consider how these new tools and techniques are being applied and whether they have applications beyond disease prevention, diagnosis, and treatment. Moreover, it is important to explore how the answers are being interpreted and publicized. The bigger question is, What are the potential risks associated with the notion that everything that has some genetic component can be quantified and even potentially engineered and improved?

CHAPTER 12

Twenty-First-Century Eugenics?

GWAS is an extremely powerful method that can deduce statistically significant links between particular SNP patterns and specific diseases or traits. It is often employed to illustrate the potential genetic basis for specific diseases, such as the risk of developing type 2 diabetes, or to assess the heritable component of quantifiable measurements known to be key biomarkers for health issues, such as high blood cholesterol levels. GWAS can be used to demonstrate a genetic basis for almost any trait or medical condition that displays a heritable component. If you have a large enough sample size, any physical attribute or propensity for disease that varies within a population with some degree of heritability can potentially be assessed by GWAS. Techniques like GWAS are also being applied by some researchers to large populations to ask some potentially disturbing questions, particularly regarding the heritability of intelligence.

Although there is nothing inherently inappropriate about this work, racist ideologues have long searched for scientific evidence to bolster their spurious claims of racial or ethnic superiority. There is nothing new in the search for genetic rationales that can explain away socioeconomic disparities through seemingly inexorable biological characteristics associated with particular groups.

One perspective often underlying these efforts is a drive for those in positions of power to defend their superior circumstances by invoking predetermined biological explanations for their wealth and influence, rather than acknowledging that they benefit from privilege or good fortune.

Worldwide, the tools of genomics are being applied to study how different genes might influence intelligence, and how genetic variation might be responsible for how innately bright individuals are as well as how successful they might be in school and beyond. If we can screen preimplantation embryos in IVF for the presence or absence of specific disease-causing genetic mutations, couldn't this be applied to other considerations, too?

Nearly a hundred years ago, some of the brightest minds in the burgeoning field of human genetics were focused upon the now-disgraced field of eugenics. To bring about social change, they attempted to link together genetic tools and assumptions that were rudimentary at best and completely inaccurate at worst. The identification of certain individuals as "undesirable" and others as "superior" became divorced from cultural or economic context; the causes of one's status, problems, or privilege were solely attributed to good or bad genetic ancestry.

Ranging from selective breeding of those deemed superior human specimens to the forced sterilization of undesirables, the tools of eugenics appear abhorrent today. It is no surprise that the early proponents of eugenics inspired the Nazis, who employed even more extreme methods in their drive for global domination of their so-called master race. That being said, some of the vocabulary and perspectives associated with eugenics persist, and the potential risks involved in employing policies and actions predicated on identifying a purported genetic basis for a supposedly beneficial human trait like intelligence are extremely concerning and alarming.

One problem is more philosophical than truly biological in nature. In large part, the human species is currently removed from much of the fundamental basis for natural selection. What exactly are the selective pressures guiding human evolution at this moment? Although our modern mechanized society and the digital world, in particular, have only existed for a brief moment on an evolutionary time scale, it is intriguing to wonder whether the human species is currently outside the evolutionary forces that have driven natural selection up to this point. Although some traits such as a functioning immune system, the ability to effectively reproduce, and the capacity to efficiently metabolize nutrients remain critical to our survival, the ability to outrun a predator or physically overpower a competitor is no longer as relevant. Ironically, much of the world that we have created seems to be operating against some of these key biological characteristics, and we are now experiencing increasing incidence of autoimmune disease, infertility, and obesity. Thus, as natural external risk factors have largely dropped away, our own internal biology has emerged as a major limit to our fitness. Much of what we now deem to be desirable (wealth, status, and beauty) comes from cultural considerations rather than biological drivers.

MEASURING INTELLIGENCE

While intelligence may seem to be readily identifiable, there are some important questions that need to be asked when trying to determine levels of intelligence: What exactly is intelligence? Who gets to decide? Is it the ability to memorize facts and figures? Is it based on how well we understand complex concepts? Does it require problem-solving skills?

Although intelligence can be "measured" by tools like an IQ test and the results will show variance across a population, that does not mean that inherent intellectual ability is genetic in basis. IQ tests are not infallible. It's important to consider the author or originator of the test as there may be an inherent bias. Based on frames of reference and personal experience, how a test is interpreted must also be considered. Questions on tests that are meant to measure intelligence certainly have the potential to suffer from biased expectations and not be equally applicable to all individuals. Cultural bias is a major issue with any analysis of intelligence.

We have to ask who gets to write the questions on the test, and how different perspectives can, from critical thinking to memorization to problem solving, be integrated together.

And even if we assume that there is something that can be described as inherent intelligence and that it can be effectively quantified, there is still an extremely important factor that needs to be considered when making conclusions based upon these types of analyses: genetic-environmental covariance. What this means is how you develop—the person you become—depends heavily upon the way the world treats you. Where you were born, how you look, and your socioeconomic class all affect your personal experience. Often, the nature-versus-nurture debate is framed as a dichotomy pitting genes against the environment. But the two are linked and cannot be easily separated. The combined effects of genes and environment are generally more easily evident in traits that are obvious, such as racial characteristics like skin tone, facial features, and hair type. We make judgments and distinctions about others all the time, and much of this is likely to be unconscious and potentially based on traits that are hard to describe or quantify. This type of unconscious bias is pervasive and can affect all types of conclusions that we draw when we interact with others.

There are myriad ways that an individual's personal genetic background affects the environment to which they are exposed, how they are treated, and how they eventually think about themselves. Think about the line outside a popular dance club. Who gets let in first? The bouncers admit the people that the establishment wants to have in the club and will likely be a draw for other patrons or will spend money. Society is like this, too. You could probably do a GWAS study of people who were let into a club, compared to those who were not, and with a large enough sample come away with some statistically significant genetic variants distinguishing the two different groups. This is a very long way from saying that those who might carry one or more of those genetic markers are fundamentally unsuited to the world inside the club.

Ranging from education to health care and the criminal justice system, racial bias leads to vastly different outcomes for people with differing backgrounds. This is because the way someone looks, which is largely genetically determined, affects the way that others act toward that person. Implicit bias and privilege are real. The word *prejudice* means to prejudge. People make snap judgments about others all the time and frequently make assessments of others' intelligence, work ethic, and honesty according to criteria as trivial as appearance.

Although a large GWAS study might suggest genetic components linking SNP variants to measures of intelligence, it is largely impossible to separate them from other characteristics that are genetically determined and can affect the opportunities and advantages available to an individual. It is not that there is definitely nothing of scientific value here; it is just that what is truly being measured and linked to a genetic background is generally way too complex to support broad conclusions. As there will always be people looking for biological evidence to support racist ideologies, there is a huge responsibility that scientists must recognize to ensure their results are not taken out of context.

There have been some recent attempts to assess heritability in quantitative measures potentially linked to intelligence that seem to be successfully avoiding common problems with oversimplified measures of intelligence, or overinterpreted conclusions. However, it seems that the interpretations of these analyses are at risk of becoming significantly overblown once they enter the popular press. Recently, major media outlets have been making breathless claims about the genetic basis for intelligence assessed through large-scale GWAS studies. Through a very large sample of over one million individuals, specific SNP patterns were identified that correlated with certain variables associated with intelligence, such as "educational attainment," how long a person stays in school, as well as more specific measures of cognitive function. There are, of course, many other facets to these overall variables such as prenatal and early childhood nutrition, socioeconomic status, and many other influences that could also be linked directly or indirectly to genetic background. Thus, pinpointing a genetic basis for intelligence comes with several critically important caveats that tend to be overlooked when faced with bold general conclusions.

As mentioned, the study described above assessed over one million individuals. A pool that large generates a huge amount of data, and even with that powerful an analysis, only about 10 percent of the variance in these measurements among the study population was accounted for in their findings, meaning that about 90 percent of the measured differences in intelligence, or related variables, had nothing to do with the genetic variants they identified. The individuals who were included in this analysis were also relatively genetically homogenous—nearly all white people. When the researchers attempted to apply a similar strategy to samples from African Americans, the results were much less statistically sound than when people with European backgrounds were assessed. The overall applicability of the specific genetic contributions to intelligence that they identified are an open question. This is particularly relevant as the study attempted to provide some mechanistic insight underlying the particular genetic variants identified. Specifically, the authors reported that many of the SNPs identified as potentially playing a role in these measures of variables linked to intelligence were associated with locations of genes known to be expressed in the nervous system. The study did not specifically demonstrate any altered expression levels or functional status as a result of particular SNPs—for example, biochemical analysis of protein variants isolated from study participants—but these are extremely interesting findings, with potentially important implications nonetheless.

The implications of studies that look at the genetic basis for desirable human characteristics like intelligence raise significant questions: Are we rapidly accelerating into a future where genetic engineering and selection of preferred traits will permit those in positions of wealth and power to further reinforce their place at the top of society? Of course, no technological development or tool is inherently evil. There are many less controversial applications of GWAS; thus, not all potential developments in this area raise these types of concerns. But caution definitely needs to be taken when looking at a set of genetic variants purported to have a statistically significant effect on something like intelligence. There are many ongoing investigations in this area, and great care is required for the researchers, the media that report the results, and the general public that might not understand all the limitations in this type of work. Another way that genomic technologies are currently being applied relates to the next generation of humans and is already revolutionizing some aspects of prenatal care.

CHAPTER 13

DNA and Prenatal Genetic Testing

G enerally speaking, the health and well-being of our children is one of our greatest concerns. This responsibility begins before they are born, and never ends. Genomic analyses have long been applied to everything from testing for different genetic conditions arising from particular mutations in specific genes, such as cystic fibrosis, to attempts at predicting the odds of an individual developing complex debilitating diseases caused by genetic variants and family history—for example, cancer or diabetes. In addition to genetic diseases that run in families, chromosomal abnormalities such as fragmentations or duplications can be a major concern for prospective parents.

CHROMOSOMAL ABNORMALITIES

Aneuploidy is the term used to describe the inheritance of more than forty-six chromosomes. Remember that each parent is responsible for twenty-three chromosomes. But when an extra chromosome is added during fertilization, every cell of the resulting child will display this extra chromosome. This could happen if during formation of a sperm or egg, one were to end up with twenty-four chromosomes, instead of the usual twenty-three. In Down syndrome, a common example of aneuploidy, the genome contains three copies of chromosome 21, instead of the usual two.

This is referred to as a trisomy, a specific type of aneuploidy. Down syndrome is also known as trisomy 21.

Unlike most dominant and recessive genetic diseases, chromosomal abnormalities are generally not believed to be inherited. Although the odds of having a child with an issue such as Down syndrome do seem to increase with parental age, there is really no way to predict whether a child might have this kind of chromosomal issue before conception. Thus, multiple tests have been developed to screen in utero for chromosomal abnormities such as the extra chromosome found in Down syndrome.

PRENATAL TESTING

There are a number of methods that can be used to assess whether a fetus has a chromosomal disorder. Analysis of detailed ultrasound images (nuchal translucency screening) can be used at certain stages of fetal development to screen for Down syndrome, but it is not the most accurate test. It does not catch a significant proportion of cases where a baby will have Down syndrome, and also sometimes reports false positives. There are also some blood tests that can be performed on maternal blood in an attempt to determine if the fetus might have specific issues—for example, neurological disorders like spina bifida. However, these noninvasive screening tools do not have adequate predictive value.

The gold standard for prenatal testing is to obtain cells from the fetus and look for chromosomal abnormalities directly from that sample. Historically, this has been performed through amniocentesis, which involves using a needle to extract a sample of the amniotic fluid surrounding the fetus. This is obviously an invasive procedure and carries a significant risk of spontaneous loss of pregnancy. Although occurring in less than 1 percent of cases, miscarriage following amniocentesis is clearly a terrible outcome that should be avoided if at all possible.

Similarly, another procedure known as chorionic villus sampling analyzes fetal tissue rather than amniotic fluid. However, this carries roughly similar risks of spontaneous miscarriage. Because of the potential risk, many obstetricians have relied on nuchal translucency and other noninvasive tests for lower-risk pregnancies, and reserved amniocentesis for those at higher risk for having children with chromosomal abnormalities, such as in cases of advanced maternal age (thirty-five years and above).

More recently, developments in genomic-based screening tests have provided a noninvasive and low-risk option to screen for genetic abnormalities, with essentially ironclad predictive value approaching 100 percent accuracy if properly performed with good-quality samples. The discovery approximately sixty years ago that fetal DNA is present in the bloodstream of a pregnant mother has led to revolutionary developments in prenatal screening. As these techniques are safer and simpler to perform than amniocentesis, they are improving health equity as they can be made available to more women and do not require specialized expert medical capabilities, and are therefore poised to replace more invasive and dangerous tests.

Referred to as noninvasive prenatal testing (NIPT), genomic analysis of small samples of circulating fetal DNA that are found in maternal blood is rapidly becoming the standard approach for most prenatal screening of this type. By taking a simple maternal blood sample and isolating circulating fetal DNA for testing, the fetus can be screened for chromosomal abnormalities such as Down syndrome and several others without the risks associated with invasive tests. NIPT can be performed as early as ten weeks into pregnancy, whereas amniocentesis is sometimes not performed until twenty weeks of gestation. This delay can further complicate the consideration of options in the case of a positive test. That being said, to be completely sure of the in utero diagnosis, amniocentesis is generally conducted for confirmation if the NIPT suggests chromosomal abnormalities.

It is important to note that screening is not technically diagnostic. The difference is that screening can suggest an increased risk or likelihood of a condition or disease, rather than identifying it with complete certainty. Think of home pregnancy tests. A negative home pregnancy test, when a woman is pregnant, could be the result of misuse, or a faulty testing kit. A positive test can also be inaccurate, so positive screens are generally followed up with more definitive tests conducted under medical supervision. Most screens carry some risk of what is referred to as a false negative, where the individual screened in fact has the condition being tested but the screen came back negative. The complete evaluation of the relative benefits and limitations of different screening procedures must consequently include analysis of accuracy (e.g., false negatives and false positives) in addition to the cost of the test and risks involved.

Initially, NIPT was primarily applied to higher-risk pregnancies. However, compared to conventional options, this testing is now being widely expanded because the potential to gain valuable information far outweighs the relative risks. Millions of women have already undergone NIPT, although it should be mentioned that it requires some genetic counseling to understand the results, and there are laboratory costs to consider. NIPT is not always covered by health insurance and can cost up to a few thousand dollars if an individual has to completely pay out of pocket. As with most genetic testing, NIPT also can provide only information for what is specifically analyzed, so besides the more common abnormalities such as Down syndrome, negative results can't be interpreted as definitive proof that no other chromosomal problems will be present. Although NIPT tests for trisomy 21, Down syndrome, trisomy 18, Edwards syndrome, trisomy 13, Patau syndrome, and extra or missing X or Y sex chromosomes, other rarer chromosomal anomalies won't necessarily be included.

It is quite easy to determine if fetal DNA is present in a maternal blood sample because fetal DNA is significantly different from maternal DNA. The primary issue with NIPT is that sometimes not enough fetal DNA is present in the maternal blood sample to be successfully tested. This seems to be more common among obese mothers. Lack of fetal DNA won't generally pose a problem as far as the accuracy of a screen are concerned, though, since the screen simply will not be performed if not enough fetal DNA is collected.

There are other concerns with NIPT, too. There can be situations that could lead to difficulty in identifying fetal DNA. Health issues in the mother can hinder the ability to perform NIPT.

Because cells derived from tumors can have extremely significant differences at a genomic level from the rest of the individual, including chromosomal abnormalities, if the mother has cancer, particularly if it is undiagnosed, it can be difficult to differentiate between tumor-derived DNA and fetal DNA. In addition, a twin pregnancy occasionally only results in the complete development and birth of a single child. In these cases of a "vanished twin," there can be circulating fetal DNA that will not correspond to the viable fetus being screened. This could lead to confusing or incorrect results in a screening test.

There are also concerns about using circulating fetal DNA for screening of nonmedical issues. NIPT can be conducted at any time after nine weeks of pregnancy, which is several weeks before the sex of the fetus can be determined by ultrasound—the conventional approach. There are concerns that early sex determination by NIPT might allow decisions of whether to terminate a pregnancy to be based upon criteria such as the sex of the child. Of course, as our understanding of the genetics of complex traits grows, and the ability to screen for particular characteristics becomes more advanced, NIPT could be used further beyond the scope it was developed to address. With the appropriate genomic analysis of fetal DNA, a great deal of other information could potentially be obtained. The use of NIPT and abortion to select against particular traits or characteristics (e.g., intelligence, beauty, or physical prowess) seems monstrous but is certainly a theoretical possibility. Sex-selective abortion is a very contentious topic. There are countries, such as India, where it is illegal to determine the sex of a fetus before birth in order to try to prevent this type of practice. As with many developments in genomic technologies, NIPT has the potential to greatly ease suffering but can also be employed for questionable or unethical ends.

NIPT is not the only potentially useful application for circulating DNA. Stephen Quake, a professor at Stanford and copresident of the Chan Zuckerberg Bio Hub, has spent much of the last decade developing techniques and applications aimed at expanding the potential impact of analysis of circulating DNA. Quake is also an inventor and entrepreneur who has a long history of developing commercial and translational medical applications for bioscience innovations—essentially taking research and building on it to develop medical tests or procedures. His lab published some of the first work demonstrating that NIPT could become a clinical reality, benefiting millions of people. More recently, he has begun to apply some of these same techniques to other medically relevant applications like analyzing tumor DNA or that derived from an organ after transplant.

Since solid tumors often shed tumor DNA into the bloodstream, sequencing circulating tumor cell DNA over time may help us develop a better understanding of tumor biology and cancer progression. Similarly, people who have undergone organ transplants will have someone else's DNA inside their body. Organ rejection after transplantation is a major issue; by analyzing circulating DNA derived from the organ donor, insights into transplantation rejection can be gained, as the appearance of donor DNA in the bloodstream of a recipient is an indicator of rejection.

The potential application of genomic technologies to medicine includes many contributions to different areas, and screening and diagnosis are just two. Treatment selection and the field of pharmacology are some other areas of medicine where genetic information is becoming increasingly influential.

CHAPTER 14

Pharmacogenomics

The field of pharmacogenomics lies at the forefront of prevalent applications of DNA sequence data to health and medicine. In essence, pharmacogenomics refers to using genetic information to assist in selecting specific pharmaceutical interventions.

It has long been known that the tolerance and efficacy of particular medications vary among people. Ranging from relative activity to the potential for toxicity, different drugs function differently depending on the patient. In many cases, these differences are dependent upon the specific genes of the patient. This means that after analysis of genomic DNA, appropriate medications can be selected, and inappropriate drugs can be avoided. Generally, toxic effects follow incomplete enzymatic breakdown of drugs, so by analyzing the activity of these enzymes, or simply by looking for particular gene variants known to make the enzymes encoded less functional, specific treatment insights can be gained.

Many drugs are given in a functionally inactive form that might be more stable for long-term storage. For example, codeine in commonly prescribed painkillers and cough suppressants must be converted into morphine by the enzyme CYP2D6, which is expressed primarily in the liver, to achieve its full effect. People with a mutant variant of *CYP2D6* that display decreased enzyme activity will not efficiently convert the codeine to morphine and should not be prescribed this drug. A more dangerous situation arises, however, for people who have inherited three functional copies of *CYP2D6* via a heritable gene duplication event. For such people, codeine is converted to morphine much more rapidly and efficiently than normal—like taking a shot, instead of sipping a beer—leading to a very rapid burst of a drug that can result in dangerous symptoms associated with overdose. If a person's specific variants of *CYP2D6* can be detected through genomic analyses—for example, by DNA sequencing—doctors can avoid prescribing that person codeine.

Other drugs must be enzymatically metabolized, and if they are not efficiently broken down, toxicity can occur owing to inappropriate buildup in the body. Variants of these types of activation and clearance enzymes are key targets for pharmacogenomics analyses.

PHARMACOKINETICS

Much of the research in the area of pharmacogenomics focuses on the general field of pharmacokinetics, which essentially concerns the different aspects of drug function besides the actual mode of action for the medicinal effect. In general, pharmacokinetics is described by the acronym *ADME* (absorption, distribution, metabolism, and excretion). Any variations in genes that encode proteins that play a role in any of these steps can result in differences in the pharmacokinetics of a particular drug in an individual. These could include enzymes involved in the activation or breakdown of a drug, or transporters that move the drug between two cellular compartments.

When considering the potential effect of any drug, pharmacokinetics is generally combined with the concept of pharmacodynamics. Pharmacodynamics is defined by the actual mode of action of the drug—what it actually does on a molecular level; for example, an inhibitor blocks the function of a particular enzyme. Pharmacodynamics can certainly be altered by individual genomic variation. You can picture an enzyme inhibitor that must fit into the target enzyme as a cork going into a bottle, and thereby preventing any wine from spilling out. But what if your particular bottle is a smaller size, and the cork won't fit? An analogous inhibitor would be unable to stop the normal function of the enzyme because of a specific structural difference in the target resulting from genetic variation.

In some cases, as with codeine, people with reduced function of enzymes involved in drug activation simply cannot feel the benefits of that medicine. In other cases, certain genetic variants can cause significant toxicity when a person is treated with a specific drug. The drug 6-mercaptopurine (6-MP) is an immunosuppressant classically prescribed as frontline therapy in autoimmune conditions such as Crohn's disease—a type of inflammatory bowel disease. The enzyme thiopurine S-methyltransferase inactivates 6-MP and is required to maintain a functional, but nontoxic, level of 6-MP in the body. This is like running water into a bath with the drain open so that some water remains but the bath doesn't overflow; whereas if the drain is plugged, the bath will overflow eventually. A small percentage of people have reduced function of this enzyme. In these individuals the 6-MP can build up and cause significant toxicity, so it is standard practice to ensure that a patient has sufficient thiopurine S-methyltransferase before prescribing 6-MP. The standard approach to doing this involves specific enzyme activity testing in a dedicated biochemistry laboratory. Alternatively, by sequencing the gene encoding thiopurine S-methyltransferase, any variants that might alter function can be detected.

The ultimate promise of pharmacogenomics is that an individual's genomic DNA sequence can be analyzed for mutations that could be relevant to any treatments that person might ever receive. Rather than incremental evaluations for enzyme activity, or analysis of individual gene sequences, all that would need to occur would be a comparison of the person's genome with known variants associated with particular drug interactions.

The potential utility of pharmacogenomics is limited only by the known connections between specific gene variants and particular drugs. With a whole genome sequence, future comparisons could be performed automatically as new potentially relevant functional variants are discovered, and the treatment needs for future conditions emerge. However, it isn't strictly necessary for the malfunction of a specific gene variant to be understood at a mechanistic level so long as a statistically significant association of a given variant has been identified and correlated to adverse outcomes with a particular drug treatment. If you know that a particular variant in an enzyme gene is resistant to a specific inhibitor through genomic analyses comparing people who respond to the drug to people who do not, it might not be necessary to perform biochemical analyses of the mutant enzyme to determine the root cause of the issue, especially if there are alternative therapies available. There are already hundreds of potential risks printed on drug labels related to specific gene variants. Of course, that does not mean that all doctors are appropriately screening their patients for potential risks before prescribing these drugs.

As with many aspects of genomic technologies, the appropriate application and interpretation of pharmacogenomics tools depends upon cost-benefit analyses, especially compared to historically accepted approaches, such as ways of testing the biochemical activity of enzymes in a test tube. Furthermore, among patients, doctors, pharmacists, and genetic counselors, the division of labor and burden of responsibility is currently not clear. This is especially so within the United States, where the healthcare system, costs, benefits, and bureaucratic organization clearly lead to complications when trying to consider these types of complex procedures linking genetic analyses, medical interventions, and drug selection and dosing. Finally, the short- and long-term economics of pharmacogenomics are being worked out as the cost and efficiency of DNA sequencing is rapidly changing and databases are swelling with correlative information.

All that being said, it is important to consider the true costs of prescribing drugs that don't work, or misprescribing drugs to people who cannot tolerate them. Some estimates suggest that there is up to a 50 percent failure rate for drugs, and that $300 billion per year might be lost because of the conventional medical approach. Doctors seem to primarily employ simple trial and error when it comes to prescribing drugs. Physicians look at the suspected diagnosis and then match it to a potentially appropriate drug that doesn't carry specific added risks based on other drugs the patient is taking, or other known risk factors. However, widespread application of pharmacogenomics could lead to greatly increased overall efficiency and efficacy, both improving the ability to treat disease, and reducing the potential for negative side effects. Ultimately, as therapies get more specific, pharmacogenomics also has the potential to serve as one arm of the currently expanding area of personalized medicine.

CHAPTER 15

Personalized Medicine

Personalized medicine draws on the understanding of your unique medical issues to tailor treatment options best suited to you. Although genomics is a key player here, the tools and technologies underlying personalized medicine are not necessarily restricted to that field. When a particular therapeutic plan is developed with specific characteristics of the patient in mind, that is personalized medicine at work. Some considerations, like drugs that cannot be combined together or therapeutic options that must be excluded—for example, in someone with a weakened immune system—might be considered standard practice, and not fit a strict definition of personalized medicine. As with the genetic basis of human disease itself, one cornerstone of personalized medicine is that although some relevant variation among individuals might be the result of single significant mutations in individual genes; especially with complex diseases and characteristics, multiple small differences, many polymorphic variants, combine to result in significant changes.

The role of genomics in personalized medicine ranges from large-scale assessment of SNP patterns that are correlated to complex diseases, to evaluating individual causative mutations potentially underlying monogenic diseases (controlled by a single gene), to sequencing panels of variants involved in heterogeneous (multiple genes and/or causes) conditions such as cancer, where multiple mutations can emerge over time and drive the disease in a vicious cycle in which mutations can continue to build up as the disease progresses. Often cancer cells undergo uncontrolled replication, and the constant copying of the genome allows errors to multiply. Furthermore, cancer cells often display mutations in the very genes that correct these errors, in addition to those that drive protective cells displaying aberrant behaviors into programmed cell death.

Personalized medicine faces several challenges: knowing which specific gene variants to test for, having genomic tools with the sensitivity and specificity to reliably measure these differences, and having the power to extract significant conclusions from this information. When many subtle differences combine to generate complex multifaceted disease states, it's critical to consider how much focus to put on understanding the mechanisms underlying the effect of each and every different gene variant relative to the practicality of simply looking for a statistically significant result—basically, a correlation rather than a causation. This is at heart a question of whether you treat the symptoms, which can occur for many reasons, or the root cause. A headache can be caused by a stressful day, whacking yourself against a door, a tumor, or a stroke. Knowing which is responsible makes all the difference. Personalized medicine simply takes this perspective to a molecular level of detail.

As with any medical paradigm, the tools and techniques that permit information gathering for research purposes can be repurposed for diagnostic and therapeutic use, although it can often take years of refinement and regulatory approval to transition knowledge gained from research into actual clinical practice.

Ultimately, however, what makes an individual unique, from a medical perspective, can be measured and quantitatively described in a manner that can be useful in a clinical setting. These assessments could take place at many levels, including analyzing and evaluating genomic DNA, the sequences of important enzymes, anatomy, physiology, and behavior. However, the current discussion will focus on genomic personalized medicine, since our genomes are the quantifiable starting point for what makes us individuals. As with any nontraditional medical approach, understanding the potential value of genomic personalized medicine must include cost-benefit analyses, as well as ethical, legal, and regulatory considerations.

Because personalized medicine makes specific measurements particular to an individual, whether it is based upon genomic DNA or any other variable, this raises the question of what a disease truly is, and whether that definition depends on symptoms or other diagnostic criteria. Alternatively, medical conditions can be defined by the underlying molecular mechanisms, which originate to large extent at a genomic level. You can tell the mechanic that your car won't start, but without knowing if it is the alternator, the battery, or the starter itself, it won't be possible to fix the problem. But if the mechanic knows that the wire connecting the battery to the starter of a vehicle of a specific make, model, and year tends to wear out quickly, the problem can be fixed and the wire swapped out before it fails without having to resort to extensive tests.

Although screening and prevention can either be viewed as distinct or linked to diagnosis and treatment, it is important not to get bogged down in semantics, though these distinctions may be useful to keep in the back of one's mind. Ideally, genomic personalized medicine can be applied to understand different root causes for disease as determined by distinct genetic variations that can demonstrate similar or even indistinguishable symptoms yet require alternative therapeutic interventions. Rather than treat all similar cases the same—for example, giving all people with lung cancer the same chemotherapy to indiscriminately kill the dividing cells—you can determine the genetic mutation that is driving the tumor to grow, possibly a mutation in an enzyme that drives cell division, and deploy an inhibitor targeted specifically to that mutant enzyme.

Personalized medicine, in many ways, represents the ultimate promise of the potential for genomics to have a direct, positive impact on human health. Although often referred to interchangeably with the term *precision medicine*, personalized medicine can be thought of as specifically tailored to the unique individual, whereas precision medicine could be slightly broader and refer to small groups of patients sharing essentially all relevant variable genetic parameters. Having a special treatment plan for all people with a particular type of tumor is still precision medicine, but it can become personal medicine when you actually conduct an individual analysis of each patient to determine a specific medical strategy.

The recent case of a new drug called milasen developed for an individual patient, a little girl named Mila, represents an extreme example of personalized medicine in action. In this case, a unique combination of mutations not previously observed that caused the otherwise fatal neurological condition known as Batten disease required development of a novel therapeutic intervention. This was made possible with funding through charitable fundraising and the therapy underwent rapid development and approval to be of use to the individual in question. This extreme example aside, taken together, the primarily semantic differences among genetic medicine, genomic medicine, personalized medicine, and precision medicine are not as important as the general understanding that the impact of living in a postgenomic world has on biomedical research, disease prevention, screening, diagnosis, and treatment.

It is very easy to consider personalized medicine as simply the result of living in a world where everyone can have their complete genome sequenced, but in practice it is much more complex. It generally holds that the sheer volume of information contained in the human genome is not as useful for medical professionals treating individual patients compared to more targeted genetic tests such as searching for particular known variants because our practical, mechanistic knowledge of biomedicine is far outpaced by all of the potential information available in the genome. Given all the differences in any unique genome relative to the general population, and how few of these polymorphic variants might actually be biomedically relevant, it can be a major challenge to isolate the signal from the noise. Of course, it is critical to understand which standard or reference an individual genome will be compared against, as there really is no single gold standard, especially considering variants that can be different from the general population and yet run in families, or are consistent across those who share a common ancestry. This is especially true given that variants associated with disease in one population might be irrelevant in another because of the potential for redundant or compensatory mechanisms, such as multiple similar enzymes that can potentially functionally substitute for each other. It is not really possible to apply a global analysis of an individual unique genome against a standard reference, as there is no one definition of what it means to be human. Although genomes corresponding to people with or without specific traits of medical conditions can be curated, and commonalities among those of shared ancestry can be cataloged, we are all individual collections of polymorphisms, variants, and mutations that come together to form a unique person. In many cases, it makes more sense to ask specific and direct questions based upon measures that we do have, from broad statistical correlations of global SNP patterns to functional alterations in specific protein structures and enzyme kinetics based upon mutations in specific genes.

There are many examples of personalized genomic medicine that focus on particular individual variations in specific gene products and target these mutant proteins with expressly designed pharmaceutical agents that selectively inhibit this aberrant activity. Cancer can often arise from mutations that increase function in proteins encoded by so-called oncogenes. In many cases, these are enzymes involved in cell replication that become hyperactive because of particular sequence changes. Drugs that selectively inhibit these mutant enzymes, but do not interact with the normal version, have been developed to combat the cancer. Developing drugs to specifically target mutated proteins causing uncontrolled cell division is much easier said than done, and successes in this area have been few and far between.

That being said, even what successes there have been have sometimes been only temporary at best because a tumor is generally a relatively heterogeneous combination of cells carrying many different mutations, and targeting only one of these specific "driver" mutations that is critical to uncontrolled cell replication might not ultimately be enough for full, complete, lifelong remission. Any cells remaining might harbor alternative mutations that eventually lead to recurrence of the tumor. Unless we test for every possible driver mutation and deploy drugs that can specifically inhibit each one, this piecemeal targeted approach has its limitations.

The drug vemurafenib was developed to specifically block the growth of a particular mutant variant of the Raf protein found in some particularly dangerous cases of melanoma skin cancer. Although vemurafenib definitely inhibits the mutant Raf enzyme and this generally leads to positive clinical outcomes, the resulting remission can be short-lived as other cells harboring alternative mutations emerge. Even more concerning is a recent report suggesting that many drugs thought to kill cancer cells through functioning as specific enzyme inhibitors might in actuality simply be exerting so-called off-target toxic effects. This seems to be attributable to errors in prior studies that aimed to identify specific mutant enzymes critical to tumor growth, the drug targets, and the understanding of the mode of action of the different chemical agents developed.

Although some drugs have limitations in their efficacy, there are also many successful examples of targeted cancer therapies developed for use in the context of particular genomic issues. The so-called Philadelphia chromosome (named for the city in which it was discovered) arises from a translocation event (an unusual arrangement of chromosomes) between chromosomes 9 and 22, and is found in many cases of leukemia (white blood cell cancer). This results in the formation of a mutant variant of the Abelson oncogene *ABL* that is genetically fused with portions of the breakpoint cluster region protein (BCR), and expressed as a hyperactive enzyme known as BCR-ABL. The drug imatinib was developed nearly twenty years ago to specifically inhibit this aberrant mutant form of the enzyme underlying these cases of leukemia. Although extremely effective and a clear success story, there are some BRC-ABL mutations that are resistant to imatinib and additional, more specific drugs based on the patient's individual relevant genetic background are being developed.

Clearly, personalized therapies have the potential to be more specifically targeted than conventional treatments, and hopefully with fewer side effects. This is particularly relevant in the case of cancer when the conventional treatments often include broadly cytotoxic chemotherapeutic agents, essentially poisons that kill dividing cells and don't necessarily discriminate between the rapidly dividing cells of a tumor and the normally dividing cells of the body. Overall cost-benefit analyses are particularly complicated considering the potential for emergent drug resistance, especially among aggressive tumors. Balancing therapeutic efficacy, both short and long term, with potential toxicity, unintended side effects, cost (both of drug development and to the patient), and integration into existing medical systems can be extremely challenging. Furthermore, the costs associated with making the appropriate genetic diagnosis based on the specific mutations underlying the disease must be considered, from both laboratory and analytical perspectives. Since there is often no fixed, objective measure to work from, today's variant of unknown significance can well become tomorrow's clinically relevant mutation.

Without therapies specifically targeting a mutant protein responsible for a disease, an alternative perspective of personalized medicine focuses upon predicting outcomes for disease. Different molecular signatures underlying otherwise identical diseases can be associated with vastly different disease progression and outcomes. If there aren't specific therapeutic options that can be deployed depending upon the results of genomic analyses, the doctor and patient can at least be armed with information relevant to disease progression, preparing them for what might likely come next, whether it be remission or a specific phase. Ultimately, the more that can be understood about the mechanisms underlying a disease, the better the chance for developing focused treatments that might be effective in the long term.

BOTTLENECK HYPOTHESIS

One major change in perspective in the use of genomic technologies has recently begun to take place based on analyses of the molecular pathways underlying normal and aberrant cell proliferation. Ultimately, the triggers and mechanisms behind cell division are very similar between normal and cancerous cells. While there can be multiple complex interconnected ways to initiate cell division through the factors that initiate the process, such as the different growth factor receptors at the cell surface, at the end of the pathway the actual molecular machinery performing the work to make one cell divide in two is relatively simple— duplicate the DNA and then split the cell in half, with one set of chromosomes in each of the two new cells—at least compared to the complicated networks of signaling pathways that control whether the cell enters into the process of division. One way to potentially treat cancer with greater success may be to specifically target the points at which triggers meet the machinery actually mediating cell division.

Initially proposed by Andrea Califano of Columbia University, this new perspective specifically highlights the importance of what he refers to as master regulators that function at so-called bottlenecks in pathways leading to cell division and, if unchecked, tumor formation. Normally, cells proliferate by responding to certain extracellular stimuli released by other cells, referred to as growth factors or mitogens (because they drive mitosis, cell division), by activating a specific series of enzymes in what is referred to as a signal transduction cascade. If you get an infection, there are cells in your immune system that sense the presence of the foreign pathogen that are separate from the cells that go around killing infected cells. The former can secrete proteins that tell the latter to undergo cell division so that an active army can be rapidly mobilized. These cells can thus be thought of as army recruiters who increase the ranks when needed, or generals who decide which specific divisions need more soldiers. The secreted factors released then bind receptors on the surface of other cells, which initiates the signal transduction pathway. This is like a long line of enzymes where one activates the next and so on down the line. The ultimate result of this chain reaction will be to increase the expression and activity of the proteins that actually cause cell division to take place. Mutations at many different places in this line of enzymes therefore can lead to inappropriate cell division— in other words, cancer.

Califano's hypothesis, which has generally turned out to be accurate, is that although many driver mutations can increase the response to these different pathways, there will be only a few master regulators at the end of the line that will receive these signals and then serve as a decision point, or bottleneck, from which the actual cell proliferation activity will originate. In many cases, it seems that the master regulators in the context of cancer are so-called transcription factors that regulate the expression levels of the proteins that make up the machinery responsible for cell proliferation. Once activated, transcription factors enter the nucleus, where they bind specifically to stretches of chromosomal DNA (e.g., enhancer elements and promoter regions that serve as binding sites for transcription factor proteins) and drive expression of particular genes. Imagine a bucket chain going from a water source to a fire. Any individual along the line could drop the bucket, but it is ultimately the responsibility of the last person in line to actually toss the water on the fire. This is the "bottleneck," the step actually responsible for the ultimate outcome.

It is possible that by focusing on these master regulators and bottlenecks, regardless of the specific driver mutation, permanent remission might be achieved. However, as the master regulators at the bottlenecks are generally receiving signals from the upstream drivers that control activation level and ultimately the downstream functions, whether aberrant or normal, these master regulators are generally not what is mutated in disease. This makes identifying and targeting master regulators involved in specific diseases challenging. So the failure of vemurafenib to permanently treat melanoma through focusing specifically on the driver mutation in Raf demonstrates the potential benefit of targeting master regulators instead.

One way of looking at this hypothesis is extending the analogy of war. In the United States war can only be officially declared through an act of Congress. The president can order specific military action directed by the heads of each military branch, and from the intelligence apparatus to the department of defense different types of activities can be carried out similar to those that occur in times of war. The Civil War, World War I, and World

War II were officially sanctioned wars declared by the United States. In this example, Congress would be the bottleneck and the members of Congress the master regulators, even if the soldiers responsible for engaging in combat perform many of the same functions regardless of whether war is officially declared. This is similar to the way in which the machinery carrying out cell division behaves during controlled cell proliferation and in the uncontrolled growth seen in cancer.

Furthermore, whether the upstream trigger was the assassination of Archduke Franz Ferdinand or Japan's attack on Pearl Harbor, these drivers must be regulated through the same bottleneck of Congress. Similarly, in the bottleneck model for cancer, regardless of the specific driver mutation, the bottleneck step governed by the master regulators determines whether cancer will result and should therefore be the focus of analytical methods and therapeutic targeting to ensure the best chance at long-term therapeutic effectiveness. This would be like convincing members of Congress not to declare war, regardless of the events that might potentially trigger it.

One further aspect of Califano's hypothesis is that different tumor types have specific master regulators. Thus, a specific type of tumor will involve a limited number of master regulators controlling the expression and activity of the proteins responsible for cell division in that tissue, regardless of which of the many upstream driver mutations might be causing the initial dysregulation. However, cancer as a fully realized and potentially deadly disease does not simply rely upon cell division. Metastatic disease requires recruiting new blood vessels, resisting normal cell death mechanisms, dedifferentiating (losing tissue-specific characteristics, forming a tumor that is structurally distinct from the surrounding tissue), and active migration of cells from the primary tumor to other locations where secondary tumors can form. Like the individual branches of the military in a war, these different cellular functions must be controlled by master regulators in cancer, as the joint chiefs of staff control the US military. An even broader analogy in the context of world war would be bodies such as NATO and the UN. It makes sense that different transcription factors that are tissue specific or control expression of particular groups of genes associated with these different functions would be the tumor-specific master regulators responsible for different aspects of the overall disease.

Unfortunately, transcription factors are notoriously difficult drug targets as they are usually present in the nucleus when functioning, making them hard to access, and generally transcription factors are not specific to one particular gene but act on many genes. It is therefore important to consider how the tools of genomics could be employed in the drive to develop personalized medical interventions. A key facet to this type of approach are the specific genomic technologies fueling these investigations.

The Human Genome Project took over a decade and approximately $3 billion to generate the first complete human genome reference sequence, combined from a number of individual volunteers. Now your genome can be sequenced in about a day for less than $1,000. Although sequencing costs have been rapidly decreasing, it must be remembered that in the case of cancer, multiple genomes can exist within the patient simultaneously, especially as the disease progresses and more mutations build up in different tumor cells. Knowing which cells to sequence over time is a critical consideration—from initial diagnosis through treatment, and potentially recurrence. In many cases, sequencing from biopsy samples is certainly feasible, although that generally requires availability of enough of the relevant tissue. As in the case of prenatal screening, using a blood test to sequence cell-free circulating DNA in cancer patients is showing great promise for noninvasive analyses. While the presence of cancer may be detected, however, it should be borne in mind that the location of the cancer is not revealed through this form of testing. It's also worth noting that procedures for generating large-scale genomic analyses are generally more about volume than accuracy. Focused sequencing of panels of specific target genes is usually more reliable on a nucleotide by nucleotide basis overall, but can also be relatively costlier. In addition, the more sequence data generated, the greater the cost of storage, processing, and analysis.

PERSONALIZED MEDICINE FOR TYPE 2 DIABETES

Another broad example of how personalized genomic medicine could be used to understand and treat a complex disease is type 2 diabetes, a disease that arises when the body's system for storing sugar can't keep up with demand and increased blood glucose leads to a wide variety of issues over time, ranging from problems with circulation to obesity.

This debilitating disease that affects millions has both genetic and behavioral components. Certain ethnic groups are more susceptible, and it can run in families. Diet, exercise, liver function, metabolic characteristics, and a host of other variables can affect an individual's risk for developing type 2 diabetes. One approach for the application of personalized medicine in type 2 diabetes prevention and treatment might be determining whether an individual should focus on behavioral modification or medical intervention, which drugs the person should be taking, and how they should be assessed for associated risks such as high blood pressure or kidney disease.

Each component in this multifactorial system can vary depending on the individual genomic background, family history, and ethnic background. It's far from clear whether a complex heterogeneous disease like type 2 diabetes should be classified according to patient characteristics like body mass index, therapeutic considerations such as dependence on insulin treatment, or more fundamental criteria such as genomic analyses. GWAS has been employed to identify SNP patterns associated with type 2 diabetes, and more detailed sequencing studies have found several gene variants that can contribute to disease risk to varying extents, including within particular genetic backgrounds. Although some of these specific mutations have been directly linked to specific functional changes in the proteins encoded, medications targeting these specific altered proteins have not yet been clinically deployed. Therefore, the current state of personalized genomic medicine in type 2 diabetes screening, diagnosis, classification, and treatment remains primarily the purview of research scientists and is not yet a widespread clinical reality.

The holy grail of personalized medicine could be for each of us to have our entire genome sequenced and for this information to be assessed for detailed health risk analysis. This could serve as the broadscale basis for identifying SNP patterns corresponding to specific risks of developing complex diseases, as well as potential variations in the sequence of genes involved in any number of specific conditions, either in a simple dominant or recessive mode or through complex mechanisms involving multiple different factors, ranging from chromosomal abnormalities to gene duplications.

WHOLE GENOME SEQUENCING AND HEALTH

As more information is amassed about the genetic basis of complex diseases like cancer and diabetes, whole genome sequencing (WGS) could be poised to emerge as the key reference source that will inform decisions about our health and well-being. Although it currently seems a long way off, a future where WGS serves as the instruction manual for everything from our diets to our sleep patterns, behaviors, preferences, and present and future medical statuses is within the realm of possibility. But making this dream a reality, and doing so in a way that can benefit the most people, might not be practical or sustainable. It is possible that alternatives to WGS will continue to be more cost effective and ultimately useful.

It is true that as an information-gathering exercise in a research context, WGS is an extremely rich source from which to mine new correlative associations and causative mechanisms for human disease states, and both WGS and exome sequencing are proving valuable for clinical diagnosis and treatment. The question is whether the current targeted approach for application of DNA sequence data to medicine as it is generally applied will ever be replaced by a more holistic perspective, where the genome and its mysteries have been analyzed and understood so completely that your specific genome becomes the variable and comparison to a reference genome serves as the gold standard for developing detailed insights into every relevant variant.

Given that evolution is driven by the development of new mutations in individuals and the potential spread of these variants within the population, the human genome is in reality a very slow-moving target. One recent study suggested that on average seventy-four SNPs are added per generation. The human genome is not a fixed entity. When it comes to diseases like cancer, spontaneous mutations that generally won't be heritable are what is driving the disease, often accumulating in a vicious cycle if problems with processes like DNA repair occur. The greater the dysregulation of cancer cells, the more mutations can build up in a dysfunctional positive feedback loop. This means a great deal of unpredictable variations that are not present within the genome at birth need to be understood to make complete use of WGS data in the context of personalized medicine.

Currently, trying to understand health and disease through the sequence of all the genes present in the human genome (the exome) is a bit like trying to learn a language by reading a dictionary. The *Oxford English Dictionary* includes approximately 170,000 words in current usage. Estimates vary widely, but the average vocabulary of any individual is much less than this, potentially ranging from 20,000 to 40,000 words. Considered in that light, majority of the dictionary is superfluous to be a fluent English speaker. Some antiquated words can still be found in the dictionary, although they are rarely used. These seldom-used words are a bit like redundant genes that are not functionally expressed, at least at appreciable levels. There are also words with more than one accepted spelling, and those correspond to silent polymorphisms thought to be functionally irrelevant. There are some words that are highly technical and only employed by few select groups. These represent the genes that are only expressed at certain times or places with very particular functional contexts.

It is critical that we also remember that only about 1 percent of the complete human genome represents protein coding sequences; the vast majority is made up of regulatory sequences that are important in controlling gene expression, as well as long stretches of unknown function, including what is somewhat unkindly referred to as "junk" DNA. Out of the twenty thousand or so human genes present in the genome, biologists really don't understand a great deal about most of them. There are tools that can specifically sequence only the gene sequences that are expressed, and whole exome sequencing, as this is called, can be much more cost effective and efficiently analyzed than WGS. However, many SNPs and other relevant variants within the genome exist outside the protein-coding regions, particularly in the enhancer and repressor regions that respectively increase or decrease expression of individual genes. We currently don't have a detailed functional understanding of many of the proteins encoded within the human genome, and generating this type of mechanistic biological understanding is extremely laborious. This is especially true considering that one gene can generate multiple related versions of a protein that can have distinct, but related, structural and functional characteristics. Focusing specifically upon genes present in the exome can be extremely complex.

Genes are, by definition, the portions of our genome that encode the proteins that are directly functionally relevant to human health and disease, physiology and pathology. Polymorphisms, variations, and mutations in these genes can lead to relevant differences, such as altered protein expression level, stability, localization, or function. Each amino acid in a protein is represented by a three-nucleotide codon within a gene. If you alter a nucleotide in a codon, you may change the protein sequence. But this is not always the case, as there are some amino acids that are encoded by more than one codon. These generally differ only in the third nucleotide in the series, which means that occasionally a single base change can maintain the same protein sequence. Polymorphisms that change the protein-coding sequence can cause significant differences in protein function, or no detectable change at all. One potentially disastrous example would be the introduction of a stop codon through a random mutation that would truncate a protein critical to normal cellular function.

Given that the protein-coding sequence is one long line of three-letter codons, changes that add or remove single nucleotides can be catastrophic as they will alter the entire "reading frame" of the rest of the gene sequence by putting everything after the change out of the appropriate context so that every single subsequent codon is now shifted. Imagine if you were making skewers for a barbeque and added a mushroom, then a piece of onion, and then a cherry tomato, over and over again in that particular repetitive sequence. Once the skewers came off the grill everyone at the table was meant to get a portion consisting of one mushroom, one piece of onion, and a tomato, in that specific order. If one of the tomatoes fell off into the grill, say the second one in the series, then the first person would get the correct mushroom, onion, and tomato, but the next person would get a mushroom, an onion, and then another mushroom, and the following person would get onion, tomato, and then mushroom. This is like a frame shift.

Because DNA must be converted to RNA before it is expressed in the form of a protein, even silent mutations within genes that do not alter the coding sequence of the ultimate protein would change the sequence of the RNA intermediate, and this could alter the amount of protein that gets expressed. This occurs because even minor changes in the sequence of the DNA, or the RNA encoded by the gene during the process of protein expression, could result in altered binding to the factors required for protein production, or the stability of the RNA molecule itself. That being said, it is usually the changes to coding sequence that are of primary importance when analyzing the sequence of the exome—the expressed genes.

Of course, genomic variation outside coding regions that don't directly alter gene or protein can also alter expression level, as the regions adjacent to genes are often where regulatory factors bind. Thus, a primary limitation in the utility of exome sequence data is the lack of information about these local enhancer and repressor regions. Changes to chromosome structure can have global effects on health beyond regulation or function of any one gene. Even armed "only" with an exome-focused perspective, our ability to make accurate and relevant predictions regarding human health and disease is currently quite limited.

It is not clear exactly how many protein-coding genes are contained in the human genome. Although the number of twenty thousand genes is generally accepted, the true value could actually be anywhere within about 10 percent of this number. The ability to predict if a stretch of DNA is actually a gene that will become a protein is not an exact science. Protein-coding status can be deduced, for example, by scanning for the start and stop codons that begin and end a protein sequence, as long as those are within the same reading frame—that is, that demonstrate the same three-letter codon order relative to the whole sequence. However, there are also many "pseudogenes" that are not actually expressed as functional proteins. These could be ancient vestiges of an evolutionary past that are no longer functionally required, or nonfunctional products of gene duplication events, or genes inserted by viruses that are no longer viable. However, even with the bona fide genes present in the exome, the amount of mechanistic functional information we have about each one varies a great deal.

Like humans, some genes in the genome are very famous, and others toil in obscurity. There are celebrity genes such as actin, which encodes the most abundant protein in most cells that functions as a key component of the cytoskeleton critical to cell stability, cell movement, and the organization of tissues and organs through junctions between cells that link them together, as well as muscle contraction. A quick search demonstrates that there have been over 120,000 scientific papers written regarding actin. At the other end of the spectrum are the scores of genes that are simply referred to by a chromosomal location, without any real insight into their expression level, localization, function, or relevance. Many of these might only be described through partial resemblance to other proteins with known functions, even if this only refers to a small piece of the whole structure of the protein.

Protein structure can be modular, and there are some common motifs shared between proteins that are seemingly otherwise unrelated. The protein Zif268 is so named because it contains a "zinc finger" DNA-binding domain—basically, a protein structure that holds on to a zinc ion—thus, this is a bit like referring to someone as "redheaded person in San Diego number 1,561." Finding out that you have a polymorphism in a protein that doesn't have a great deal of information available would be like hearing that "redheaded person in San Diego number 1,561" might have a cold. This is much less impactful than finding out detailed information about someone that you know a lot about—for example, that your grandmother broke her hip or that your favorite singer has throat cancer.

The volume of information provided by WGS or exome sequencing far outweighs our ability to interpret that data primarily because of limitations in the "annotation" of the genome. Understanding whether a sequence that looks like a gene is really expressed and when, where, and how this is regulated; and knowing the function of the resultant protein, with what other proteins it interacts, and how this is all regulated is critical for us to begin interpreting how changes to the sequence of the gene might affect human health and disease. One key takeaway from the application of the genome to our health is that we need a great deal more basic scientific research focused on the molecular and cellular biology that underlies human health in order to inform clinically relevant endeavors such as personalized medicine. Without a firm basis in the biological understanding of how proteins function in the context of the cell, we can't possibly hope to translate diversity at the level of the genome into medical insights.

COMMUNICATING WITH PATIENTS

One major question in the area of genomic personalized medicine is how to best communicate with patients regarding the results of large-scale analyses such as WGS, and who should be doing this communicating. Genetic counselors are well versed in discussing issues like carrier risk in the context of prenatal screening, and for those suffering from debilitating genetic diseases, such as monogenic conditions like cystic fibrosis, because the relevant genetics are relatively straightforward. Furthermore, pharmacogenomics provides information at the level of pharmaceutical interventions, such as particular drugs that should be avoided, and physicians and pharmacists are hopefully receiving training in these areas. However, the complexity and uncertainty that come along with WGS or exome sequencing data require that a different type of

mindset be taken toward patient communication and consultation. Given the sheer volume of polymorphisms present in any
individual's genome relevant to an accepted reference standard
placed in the limited context of our true mechanistic biomedical
understanding of structure and function of particular gene products, much of this diversity can be identified but not specifically
detailed in a relevant and accessible way.

How we define the border between genetics and genomics is
not merely a semantic distinction. When it comes to monogenic
diseases, the worlds of biology and medicine have been relatively
successful at integrating information from research studies into
clinical practice. Individuals can be screened with nearly absolute
certainty to determine if they are carriers for specific mutations
associated with particular genetic diseases, such as the *CFTR*
ΔF508 that causes cystic fibrosis or a premature stop codon in the
dystrophin gene *DMD* that could lead to Duchenne muscular dystrophy. The application of personal genetic testing to questions of
ancestry demonstrates how large databases can easily translate
into easily comprehensible information for the end user. However,
given the sheer volume of information provided by WGS, or even
exome sequencing, how can our health-care system possibly keep
up? Given that many of the actual insights into human health and
disease that might come from WGS or exome sequencing will be
context dependent, these results must be considered relative to
many different variables, ranging from ancestry and heritage to
behavior and environment.

How would results stemming from WGS or exome sequencing best be communicated to the general population? Beyond conventional approaches that look at specific known variants that
can either result in monogenic disease, such as the connection
between *CFTR* and cystic fibrosis, or have been correlated with
significantly increased risk of a complex disease, such as *BRCA1*
and breast cancer, what can be said about the SNPs and other variations sure to be present in anyone's specific genome? What about
the potential for errors present in these data sets stemming from
the inherent limitations in the analytical platforms? From a practical perspective, it would seem simultaneously both essential and

unfeasible to provide continuous monitoring of updates relative to relevant discoveries, but this already doesn't happen with things like prenatal screening and "variants of unknown origin." So it would seem our health-care system isn't properly set up to handle the limited analyses currently being conducted. How this—information, interpretation, updates, and application—can all be integrated into a widespread clinical reality remains to be seen.

CHILDREN'S HEALTH

If parents have reason to believe their child may have inherited gene variants carrying significant risk of a debilitating disease, it is hard to argue against testing that child through whatever means might provide the best and most detailed actionable information. It gets more complicated when you consider all the other information that could arise out of testing. Should the parent be provided the WGS and all the currently available analytical insights? What about as the child gets older and more potentially relevant information comes to light? Not everyone wants to know everything, and if the child grows up to be an adult who does not want to be bombarded with a constant barrage of ever-changing relative risk data and behavioral recommendations, will there be a way to opt out? As the power of genetic information greatly increases when multiple family members are analyzed together, other ethical and practical concerns crop up. What if siblings or children want to know about the results obtained by a relative in order to put their own health-related genomic data into context? What if they don't want to know? Without clear legal, ethical, and regulatory frameworks, this is all bound to get very complicated very quickly.

What if you find out that you are a carrier for a mutation that can cause a recessive genetic disease? Are you compelled to inform your children or your siblings? What if they don't want to know? What about more complicated and nuanced information such as an increased risk for developing a complex multigenic disease, particularly one with both genetic and behavioral components? Personal privacy must be viewed through a new lens—our genomes are unique, but at the same time they are not fully anonymous, as we share our DNA with our relatives. Furthermore, combining genomic data with information ranging from height and weight to blood pressure and other types of clinical measurements adds significant value to the DNA information but also introduces further risks to personal privacy.

How far can the benefits of combining clinical details and genealogical information to our personal genomic profiles be taken? At what point does family medicine transition into public health? What are the limits of doctor-patient confidentiality?

In the United Kingdom there has recently been a very interesting, and quite upsetting, lawsuit. A woman has sued St. George's Hospital in south London because doctors there had knowledge that she had a 50 percent chance of inheriting an incurable genetic disease and did not inform her. The woman's father was diagnosed with Huntington's, which is caused by a dominant mutation and develops later in life into a progressively debilitating neurological condition. He chose not to tell her, and she did in fact inherit the disease-causing mutant gene. Furthermore, she had a child herself. The lack of relevant information might have altered her choice of having a child who would be at risk for this terrible disease, and this case has significant implications for doctor-patient confidentiality spanning generations. At this point it seems quite impossible to imagine an ethical legal-medical framework that will simultaneously protect individual privacy, maintain doctor-patient confidentiality, and protect the right to know critical information—or not know it, if that is your preference.

Individual countries and regulatory bodies are interpreting these issues differently and, in some cases, developing incompatible recommendations and rules. Ideally, biomedical research and initiatives aimed at reducing human suffering should not be bound by national borders, and the benefits of combining genomic data from different distinct populations need to be considered. The definition of personal privacy and the types of individual data that can be stored, shared, or employed in research and clinical practice vary tremendously depending on the context. Even within the European Union, member states have differing regulations regarding the uses of genomic data sets for experimental and medical uses. No single standard governing genomic personalized medicine has emerged.

DNA BEYOND INDIVIDUAL HEALTHCARE

Direct roles in health care are not the only uses for genomic information that can raise ethical and economic concerns. Genetic testing and DNA sequencing are potential sources of profit not only for biotechnology companies but also for health insurance and medical establishments. The potential for gene therapy and genome editing to play a widespread role in health care in the future is opening up entirely new ethical questions, as well as avenues for revenue generation through novel medical interventions. Beyond the need for more genetic counselors and lab technicians, the advent of personalized medicine seems poised to create new jobs, and these inventions and innovations are already having significant economic implications. That being said, as is the case with most cutting-edge diagnostics and therapies, if we are not careful, personalized medicine might end up solely benefiting those select few with access to the best health care.

Personal genetic tests like 23andMe might serve as a helpful guide for incorporating information regarding relative risk based upon genetic analyses. Most of the people who have opted to employ these services to date have been white and well educated. This could be because of both cultural differences and financial inequalities. Although relatively affordable, personal genetic testing kits are nonessential expenses primarily catering to those with discretionary income. And what do people do when they receive their results? How does human nature influence our interpretation of information and potential changes in behavior stemming from these insights?

One study that looked at several hundred users of 23andMe who opted in for health tests had some interesting but not altogether surprising results. What the researchers found seemed to uphold a long-standing observation in the field of cognitive psychology known as the "good news–bad news asymmetry."

Basically, this means that people are more likely to change the perception of their individual risk for a medical condition with a genetic component when provided with information that suggests reduced risk compared to increased risk. It seems we are primed for good news that supports positive conclusions, rather than bad news that suggests increased concern.

When the researchers asked study participants about how they perceived their own relative risks of complex multigenic diseases before and after testing, their opinions were more likely to shift in a positive manner if they received results placing them at lower risk, than to shift in a negative manner if they received results placing them at a higher risk. In a nutshell, you might be more likely to cancel that doctor's appointment if you learn you have reduced risk for a condition than schedule a new one if you find out you have higher risk than you originally thought. This discovery calls into question how useful these tests may be if people might not seek medical intervention or follow-up when personal genetic testing suggests they are at a lower risk, as well as if it takes a disproportionately large degree of increased risk to induce significant concern.

This is a very dynamic and exciting time for personalized genomic medicine. The potential outcomes resulting from cheap and quick DNA sequencing must be balanced by weighing the distinction between information and knowledge. Assessment of potentially relevant genomic variation within the population is not as simple as other medical modalities employed for screening and diagnosis of disease. Although genome sequencing might be as cheap as getting an MRI, figuring out what to do with the deluge of resultant data is far from simple. The economic, social, and ethical implications cannot be ignored. Placing all of this information in a larger context based upon family history, ancestry, behavior, and environment makes the broad area of genomic personalized medicine extremely complex and one that is likely to go through several different phases and revolutions.

CHAPTER 16

Gene Therapy

G ene therapy is the catchall term used to describe any procedure using exogenous genetic information to treat or cure disease. There are several distinct paradigms broadly included in the field of gene therapy that can vary depending upon the exact medical issue being treated, what underlies it, and where it is found anatomically. Gene therapy is currently on the verge of a breakthrough thanks to new tools developed for genome editing drawing on techniques that, in turn, are based upon a phenomenon referred to as clustered regularly interspaced short palindromic repeats (CRISPR), which will be explored in the next chapter. The basis for human gene therapy has existed for decades and has seen many ups and downs over the years, and many periods when this seemed truly poised to become a widespread clinical reality. Therefore, before focusing on any one specific subarea of gene therapy, it makes sense to start with a more general perspective and focus on the history of this perennial "potential" disruptive medical advance.

Over the years, various gene therapy approaches have been developed and applied to specific diseases. What these different approaches all share in common is introducing foreign DNA into the people suffering from a genetic disease to cure it. Although technically RNA could be employed in similar ways, DNA is usually the means through which gene therapy is performed. Use of so-called transgenes (any DNA that didn't naturally originate in the host, target or patient) could be for temporary or permanent therapeutic effects, but most of the focus of gene therapy has been to cure genetic disease. One of the critical issues in gene therapy is how to deliver the transgene, as this needs to be done in a safe, efficient, and permanently corrective way. The so-called vectors that are used for transgene delivery in gene therapy have classically been based on viruses that can infiltrate cells and introduce exogenous DNA. The viral vector is like the missile, and the transgene the payload.

The viral vectors used for gene therapy have had the structural viral genes removed or disrupted so that the target cell won't start producing new viral particles after treatment. Furthermore, the naturally occurring parent viruses of some gene therapy vectors normally insert DNA directly into the host cell genome as part of the viral life cycle. The risk of unpredictable insertion of DNA from gene therapy into genomic DNA is a significant concern as this could result in unintended issues if it occurred within an important gene or regulatory region. Thus, the potential for uncontrolled genomic insertion of viral vectors and transgenes is a major topic of research and a significant subject for risk assessment of potential gene therapies.

There are different types of viruses that have been employed for gene therapy trials. Generally, the viruses used are significantly altered from their naturally occurring counterparts and genetically engineered to make them more suitable for use in gene therapy. Some can hold larger transgenes, and others are more efficient at introducing the transgene into cells. But significant risks are involved in the process, including side effects from immune responses to the viruses used as the vector—for example, if the patient's immune system has been primed to respond to a related virus.

Although after years of development focused on improved safety, these concerns have largely been alleviated, catastrophic adverse events with devastating personal consequences have occurred in gene therapy trials, setting back development and implementation of the field as a whole. One of the most dramatic tragic outcomes of a gene therapy trial was the untimely death nearly twenty years ago of a young man named Jesse Gelsinger during a study at the University of Pennsylvania. Jesse's death from a severe allergic reaction to the vector used for the gene therapy trial increased focus on safety in addition to efficacy, and it seems now the field is reaching maturity and promising to provide significant benefit to people suffering from diseases that can be treated at the genetic level. Although there are only a handful of human gene therapies currently approved by the FDA—such as treatments for a certain type of visual impairment that can lead to blindness and spinal muscular atrophy (SMA), which can cause fatal musculoskeletal problems in very young children—there are hundreds of pending new drug applications in this area. Zolgensma, the trade name for the gene therapy for SMA, has especially made headlines, but not necessarily in a good way. The pharmaceutical company Novartis set the price for Zolgensma at $2.1 million, which is not a price many families can afford, even for a life-saving therapy. Further complicating matters, it is not clear to what extent health insurance companies are likely to cover gene therapies.

GENE THERAPY FOR HEMOPHILIA AND OTHER DISORDERS

Hemophilia is an example of a long-standing key target in the field of gene therapy. Affecting about twenty thousand people in the United States, hemophilia is a disorder in which blood doesn't appropriately clot, and the lives of hemophiliacs can be significantly compromised in many ways. In addition to severe risk of bleeding from a cut or other injury, increased risk of internal bleeding is also a major concern for people with this disease. Many hemophiliacs have had to receive blood transfusions to replace lost blood after accidents and injuries, and this has historically carried significant risk of transmission of infectious disease. Furthermore, blood is a precious commodity that is not always available, especially in more remote locations.

Blood clotting is the result of the combined action of a series of enzymes referred to as clotting factors. There are different mutations in specific clotting factors that cause hemophilia, as well as different manifestations of hemophilia. If the problem originates in factor VIII, the disease is referred to as hemophilia A; if the mutation is in the gene for factor IX, then the individual has hemophilia B. A is more common than B, but apart from the genetic cause, the two diseases are extremely similar.

Hemophilia represents an example of a genetic disease that could essentially be cured by introducing genetic material encoding a functional replacement for the mutated gene. In contrast to other genetic diseases caused by hyperactive mutant proteins that must be inhibited to obtain a therapeutic effect (e.g., some types of cancer), hemophilia is caused by the absence of functional clotting factor that requires introduction of a replacement capable of providing the missing activity.

Conventional treatment for hemophilia includes regular infusion of clotting factor proteins, typically weekly. However, this treatment protocol is extremely costly and must be continued throughout the life of the patient. Employing gene therapy to trigger the body's own cells to generate the missing clotting factor seems like an ideal alternative. As clotting factors function in the bloodstream, the cells engineered to express the replacement clotting factor could be anything that is in direct contact with the bloodstream and can efficiently release the product of the gene therapy vector. Although up to this point successful development of gene therapy for hemophilia has been hindered by issues such as the ability to generate enough clotting factor in the blood to make a significant difference in the lives of patients, hemophilia B is emerging as a critical success story in making human gene therapy a clinical reality owing to a fortunate observation concerning another rare blood disorder.

About ten years ago a young man with a blood clot in his leg was admitted to a hospital in the city of Padua in northern Italy. There was no evidence of a clear cause, such as disease or injury. But when the inherent clotting capacity of his blood was analyzed, it turned out that although he had normal levels of factor IX in his blood, the activity of this enzyme was nearly ten times higher than average. It turns out that this individual and some of his close family members carry a mutant variant of factor IX that produces a hyperactive enzyme. This is like the Arnold Schwarzenegger or Usain Bolt of enzymes. Although this creates health risks for the affected individuals carrying this genetic variant because of increased potential for blood clots, this discovery has served as the key to developing safe and effective gene therapy for hemophilia B.

The conventional strategy for gene therapy for hemophilia B has included packaging normal factor IX into a modified virus and injecting it into the liver in an attempt to balance the permanent recovery of clotting function with limited potential for toxicity or side effects. Unfortunately, with the limited functional efficiency of normal factor IX, this has not been realistic: even with very high doses that increase the risk of unintended liver damage through hepatotoxicity, normal levels of clotting activity have been hard to obtain.

Now, however, in the wake of promising results in cellular studies and animal models, doctors have begun employing the mutant factor IX-Padua, and it has been a game changer. Recent trials with factor IX-Padua have required much lower doses—and consequently, significantly decreased risk for liver toxicity—and have resulted in much higher clotting activity. Although these trials have been able to follow patients for only a little more than a year following treatment for the first time, almost all patients receiving this therapy have been very nearly cured.

Treatments for other blood disorders are also seeing significant advances thanks to gene therapy. Thalassemia is an inherited blood disorder caused by mutations that decrease the oxygen-carrying capacity of the blood. Hemoglobin is the protein that binds to oxygen in our red blood cells, and it is formed by the combination of the products of the alpha and beta globin genes. Many people with thalassemia rely on regular blood transfusions for survival, as they express nonfunctional mutant beta globin. Recently, researchers were able to remove the stem cells that produce red blood cells from patients, treat them outside the body with a vector containing a functional beta globin gene, and then reinfuse those cells into the patient, where they produced normally functioning red blood cells. This resulted in significant benefits to all patients in the trial, either eliminating or significantly reducing their need for transfusions.

Both the factor IX-Padua study and the thalassemia report involved contributions, including funding, from biotech companies, Spark Therapeutics and Bluebird Bio, respectively. Interestingly, the viral vectors involved in both studies were the products of the industrial partners involved. This underpins both the critical importance of viral vector development for the successful implementation of gene therapy protocols and the potentially lucrative intellectual property involved.

VISION LOSS AND GENE THEREAPY

Another area in which Spark Therapeutics has made significant advances is gene therapy for blindness. Mutated genes that can cause vision loss can be functionally substituted through application of gene therapy, and the easily accessible nature of the eye for direct application of the therapy makes this a potentially noninvasive target. Leber's congenital amaurosis is a rare form of progressive vision loss caused by inheriting two mutant copies of the *RPE65* gene. Luxturna, a viral vector containing a functional copy of *RPE65* developed by Spark along with academic and medical partners, has been approved by the FDA for use through injection into the eye for treatment of this disease. Only a few thousand people are thought to suffer from Leber's congenital amaurosis worldwide, and with an expected cost of nearly $1 million for the therapy, it is not currently clear how widespread the implementation will be. Although a huge step forward for the field, and obviously a tremendous benefit to the patients who receive the treatment, this raises significant questions regarding the costs of development and clinical implementation of gene therapy and the overall applicability within our health-care system. At this stage, choices regarding the diseases to target have to balance the clinical impact—including the number of people affected and severity of the outcome—with understanding of the genetic basis and how easily it can be corrected, along with the accessibility of target cells that could have a functional impact if genetically altered.

Recent studies in animal models have suggested that gene therapy might be employed to prevent blindness in those suffering autosomal dominant retinitis pigmentosa (adRP), which is caused by mutations in the rhodopsin gene that produce a protein that is actually toxic to the light-sensing cells in which it is expressed. The mutant rhodopsin kills these cells in the eye, and over time causes blindness. In this case, the goal of the gene therapy is not to introduce another functional copy of the mutant gene but to decrease expression of the aberrant protein. To do so, researchers are using creative molecular tools that specifically silence expression of the mutant rhodopsin by targeting the RNA that encodes it for destruction in the cell and eliminating expression of the mutant protein. One caveat of this type of approach is that it cannot facilitate the recovery of sight once it is fully lost because it cannot resurrect the dead light-sensing cells. Ideally, it would need to be employed before a patient has become completely blind to slow or prevent further loss of sight.

It is clear that from anatomical accessibility for the introduction of genes to mode of action to the potential for cellular toxicity and possible immunological responses the variables governing the development and implementation of a potentially successful gene therapy approach are complex. Ethical and economic considerations regarding the target patient populations and associated costs must also be considered. After decades of little successful clinical implementation, FDA-approved gene therapy protocols are now finally becoming available. That being said, recent advances in the level of specificity and control possible in genome editing are now ushering gene therapy into an entirely new era.

PART IV

Genetic Engineering

CHAPTER 17

CRISPR–Clustered Regularly Interspaced Short Palindromic Repeats

Genetic engineering includes the ability to perform selective genomic alterations and has been used for decades to employ a wide variety of approaches at differing levels of specificity and efficiency. Foreign DNA can be injected into a cell or embryo with a tiny needle; electrical current can be employed to poke holes in cells and allow DNA from the outside to float through; tiny metal particles coated with DNA can be fired at the cell using a "gene gun"; or small bubbles of lipids containing DNA can be used to deliver DNA to a cell. If the goal is permanent alteration of the genome, once inside the cell, usually scientists will rely on relatively low-probability random insertion of foreign genes into a chromosome, or recombination events that swap an altered version of a gene for the naturally occurring one.

At many times in the past, the potential for genetic engineering to result in unwanted and unforeseen outcomes—such as spread of the genes introduced into untargeted individuals, or genomic damage from unpredictable and unintended insertions or deletions—has provoked no small amount of alarm. While these dystopias have not yet emerged, new advances in science are prompting some to entertain doubts about some of our present-day scientific goals.

Depending on whom you ask, we are either currently experiencing a revolution in the potential for genome editing that is likely to improve nearly all aspects of our lives—from our health to the food we eat—or it is now becoming readily apparent that the technique colloquially referred to as CRISPR will never live up to the hype, or could result in significant ecological damage and unintended consequences to our species and others caused by nonspecific effects on the genome such as alterations at locations other than the target.

As is the case with any disruptive technology, the promise of CRISPR (short for "clustered regularly interspaced short palindromic repeats") can be appreciated only from the point of view of potential applications, the assessment of associated risks, and the possible limitations that might restrict its usefulness. These limitations include both possibly insurmountable problems with development and implementation and regulatory restrictions related to safety and ethical concerns. Previous negative outcomes from attempts to alter the human genome in a medical setting for disease treatment in the broadly defined field of gene therapy, such as the untimely death of Jesse Gelsinger described in chapter sixteen, have understandably made some people very nervous about plans to widely deploy CRISPR-based approaches in the health-care system. Also, the news that the production of genetically modified organisms (GMOs) can potentially be performed with much greater ease and efficiency than via current means, and potentially without the same regulatory hurdles as some other types of genetic engineering, may not be welcome to all as there are some concerns about consuming GMO foods and food products as well as their impact on the environment.

EDITING THE GENOME

CRISPR has the capability to add or remove specific genes with incredible efficiency and also to edit the precise nucleotide sequence of particular target genes seemingly at will. Researchers are developing new CRISPR-based techniques, including technologies and reagents with specifically designed characteristics for targeting (the ability to make the desired alteration), specificity (avoiding any other unwanted alterations), and activity (doing so in a complete and efficient way). Furthermore, CRISPR can be applied to cells in isolation, intact organisms, plants, animals, bacteria, insects, and essentially anything with a genome.

The potential benefits of CRISPR-based techniques compared to previous methods of genome editing cannot be understated. In addition, the relative ease and flexibility of CRISPR-based approaches makes it truly superior to other techniques. The increased specificity and efficiency of CRISPR compared to other methodologies raises the question of whether CRISPR-based approaches should be classified as unique in comparison to previous technologies, or simply as extensions of a long series of developments subject to different restrictions and regulations. This is important as it will have implications for the research and development process, in addition to any ultimate clinical or commercial applications. However you frame it, CRISPR represents a potentially game-changing advance in the development of gene therapy approaches, for GMOs, and in many other targets that carry a DNA genome.

Referred to simply as CRISPR, this methodology is actually a series of related technologies and techniques, each with subtle but significant differences. The basis for CRISPR is the site-specific targeting and ability to alter particular DNA sequences. This means that an individual gene can be added, removed, or edited within the genome and have specific sequence modification introduced; disease-causing mutations could be corrected or new genes could be added in a commercial context, as in the case of genetically modified food. In nature, for example, the natural process that gave us CRISPR evolved as a defense in bacteria such as staphylococcus and streptococcus to prevent infection by so-called bacteriophages (viruses that infect bacteria).

There is a long history of studying how bacteria prevent infection by bacteriophage, and of using this knowledge for biotech purposes. One classical example is the use of restriction enzymes that can cut up foreign DNA. One of these restriction enzymes is named EcoR1, called thus because it was first isolated from *E. coli* bacteria.

Many restriction enzymes have been identified, and they all function in a similar fashion: they cut particular DNA sequences. If you add the target sequence to a piece of DNA, you can then cut it at that site whenever you want simply by adding the appropriate restriction enzyme. One of the very interesting things about how restriction enzymes work is that they leave behind overhanging ends in the target; they do not cut each DNA strand at exactly the same location. This provides the ability for the remaining overhanging ends to reattach after the cutting is done—in fact, the overhangs left over from restriction enzymes are colloquially referred to as "sticky ends." If you have two restriction sites surrounding a specific region of interest on the genome, you can remove it and the two remaining overhanging ends will stick together, re-forming a junction like a dovetail joint in carpentry. This is because, as shown in the box below, the sites that restriction enzymes cut are identical in reverse between the two DNA strands, and in this way they pair together before and after the enzyme cuts the target sequence.

CRISPR can act like this, too, only with much more flexibility and specificity. In the case of CRISPR, the bacterial hosts retain sequences matching previous bacteriophage infections, and these can then be used to recognize and disable new bacteriophages if encountered again. CRISPR involves a molecule of RNA that matches the gene sequence being targeted, a so-called guide RNA, as well as a CRISPR-associated protein (CAS), which mediates the genome editing. Other factors are involved depending on the specific method being employed. There are multiple CAS enzymes that have been identified, but the most highly publicized component is the enzyme Cas9 that cuts DNA at specific sites as dictated by the associated guide RNA. In this context, the CRISPR-Cas9 system has the ability to specifically target a particular gene for permanent disruption. This generally involves cutting the DNA sequence at that site.

Box: Function of restriction enzymes

EcoR1 target site:

GAATTC

CTTAAG

EcoR1 cutting:

| GAATTC | Region to be excised | GAATTC |
| CTTAAG | | CTTAAG |

Resulting sticky ends:

G   AATTC

CTTAA   G

Junction formation:

GAATTC

CTTAAG

However, this type of cutting is different from the way that restriction enzymes function. The CRISPR system results in blunt ends, not overhanging fragments like restriction enzymes. Blunt ends with no overlap between them can't simply rejoin. In fact, a specific mechanism referred to as blunt-end ligation needs to be employed to repair this type of event. Blunt-end ligation is generally only activated in cells that have undergone significant potentially traumatic DNA damage, such as a fracture or break in a chromosome. Thus, the cellular DNA repair responses to chromosomal breaks seem to be activated when CRISPR-Cas9 is employed. This can mean other unwanted effects, as blunt-end ligation is not always completely specific or predictable and any two free blunt ends, not necessarily just the two being targeted, can potentially be stuck together.

CRISPR-based techniques and the specific machineries involved can vary widely and are the subject of a vast amount of ongoing research. Methods have been developed for editing the specific sequence of a gene in the context of genomic DNA, increasing or decreasing the expression of specific proteins, or adding or removing particular genes. The use of CRISPR to alter specific DNA sequences within the context of a particular gene may be less globally traumatic to the genome than excision of whole genes. Some studies are suggesting that when CRISPR is employed in rapidly dividing cells, such as those of the immune system, the potential for off-target effects such as alterations at unintended genomic locations might be increased. If, for example, cutting or ligation took place at unwanted sites, this could lead to catastrophic outcomes where genes critical to survival are effectively inactivated in an unpredictable fashion.

Finally, as the bacteria that have been employed as the source for the Cas9 enzymes employed in some CRISPR applications can be human pathogens (e.g., staphylococcus and streptococcus), it is possible that the immune systems of some people might be primed to fight off the machinery of genome editing, a wrinkle that could significantly hinder proposed medical applications. Of course, as with other conventional therapies like organ transplantation, medications that reduce the function of the immune system could be employed to mitigate this issue.

It took nearly thirty years from the start of the first gene therapy–based clinical trial until the initial FDA-approved gene therapy, Luxturna from Spark Therapeutics, which (as described in the previous chapter) targets Leber's congenital amaurosis. The US systems of medical research and therapeutic development are by design extremely methodical and primarily concerned with safety. Furthermore, developing tools for DNA sequencing and genomic analysis definitely also provide the ability to assess the potential for off-target effects like unintended deletions. But it should be noted that what might on the one hand be considered off-target effects of using CRISPR might, on the other hand, in fact be normal activities of DNA repair mechanisms. This could turn into a very complicated situation to resolve. As DNA repair is critical to survival, simply employing general means to broadly inhibit these processes might not be appropriate or safe. It might be that CRISPR is less appropriate for use in certain situations—for example, in rapidly dividing cells. There is currently no consensus on how CRISPR should be limited or applied. From international groups to federal laws to funding bodies to universities and individual researchers, limits imposed upon genome editing with CRISPR, such as application to human embryos that will then carry heritable alterations in all cells, are being developed by some and avoided by others.

CRISPR-based therapies are already beginning to be developed. There are studies employing CRISPR to modify human lymphocytes outside the body to make them better able to orchestrate immunological targeting of tumor cells, or employing CRISPR to cure a genetic disease in the lungs of neonatal mice. Recently, a clinical trial was announced to assess the safety of a gene therapy for a form of Leber's congenital amaurosis that aims to employ CRISPR in the eyes of those suffering progressive vision loss. Before we can truly appreciate CRISPR's revolutionary potential, we must first look at the facts and individuals involved. In many ways, one of the most highly publicized aspects of CRISPR developments thus far has been the battle over who should hold the patent for it.

OWNERSHIP OF CRISPR

The arguments over who should own the right to develop CRISPR for commercial application have involved researchers, universities, venture capitalists, and biotech companies. An application for a CRISPR patent application based on groundbreaking proof-of-principle work performed by Erik Sontheimer was submitted a decade ago by Northwestern University, where he worked at the time. However, this appears to have been too preliminary and lacking in mechanistic detail, and the patent application was rejected. Over the next few years rapid developments among researchers working in this area paved the way for an extremely bitter dispute regarding the invention of different CRISPR-based techniques, and their potential commercial and medical applications. Although many individuals and institutions have been involved, most of the critical developments in this saga have generally involved the researchers Emmanuelle Charpentier and Jennifer Doudna, along with the University of California system, on one side, and George Church and Fred Zhang, along with the Broad Institute, Harvard, and MIT, on the other.

Critical points in the development of CRISPR for potential commercial application include a landmark 2012 paper by Charpentier and Doudna showing how CRISPR functions in what they referred to as "adaptive bacterial immunity." Understanding the normal biological mechanism and function was a major step forward for the field and paved the way for developing applications of CRISPR-based techniques beyond bacterial systems. This paper would serve as a cornerstone in the arguments made by the University of California in the patent battle to come. However, if one key paper had to be cited as most relevant to the subsequent dispute, it would probably be the 2013 report from Zhang's lab that demonstrated the first successful use of CRISPR for genome editing in human cells. Simultaneously, George Church, Zhang's colleague at Harvard, published work demonstrating "human genome editing" employing CRISPR. One greatly simplified way of framing this issue is that Charpentier and Doudna can be credited with understanding how the natural bacterial system works, but Zhang and Church have been credited with applying those tools to alternative systems beyond bacteria, such as human cells.

One potentially important consideration when trying to understand these types of patent disputes that involve individuals as well as research institutions is that in 2013 the US government transitioned from a "first-to-invent" to a "first-to-file" system. In many ways, this new approach should simplify the patent resolution process, as it now focuses primarily on exactly what is claimed in a patent filing, rather than exactly who developed what and when. Furthermore, the perspective of "potential" developments and applications doesn't seem to hold a great deal of sway in such matters, as evidenced by what transpired with Sontheimer and Northwestern University. Rather, the key consideration seems to be which patent filing contains appropriately compelling evidence for the invention in question.

Although the University of California filed a patent application based on work performed by Doudna and her collaborators before the Broad Institute filed one based upon Zhang's, the most important questions from a patent point of view focus on the experimental evidence described in each, and the potential claims of direct future development that could be supported. This raises interesting and relevant philosophical and practical questions regarding the distinction between discovery and invention, and more generally the potentially conflicting perspectives of basic versus applied research. Following challenges from the University of California in 2018, after a lengthy series of arguments, filings, and deliberations, the Broad patent, which covers most practical applications of CRISPR in plant and animal systems, was upheld by the US Patent and Trademark Office (USPTO). However, even more recently, the USPTO ruled that aspects of the Broad patent did actually interfere with the UC intellectual property, further complicating things. This is also taking place in a global landscape of other patent applications and approvals, as there have been over 12,000 CRISPR patent applications worldwide, and over 740 have been granted.

APPLICATIONS BEING EXPLORED FOR CRISPR-BASED TECHNIQUES

The potential uses of CRISPR gene editing are numerous and wide ranging. Humans have been cultivating plants and selectively breeding domesticated livestock for thousands of years. CRISPR is now being applied to dramatically accelerate these endeavors. One recent example is the ground-cherry, a relative of tomatoes that could have commercial potential if certain traits could be modified, such as increasing fruit size. Guided by knowledge gained through genomic analyses of tomatoes, scientists are currently using CRISPR gene editing to produce ground-cherries with characteristics that are more suitable for agricultural cultivation and commercial distribution and sales. Another example is a white button mushroom that had an enzyme that causes cut mushrooms to turn brown, which CRISPR-based genome editing has managed to disable.

Interestingly, this modification was approved by the US Department of Agriculture, avoiding the process required for regulating the mushroom as a GMO. The secretary of agriculture's decision, it seems, was based upon the conclusion that this same change could have been brought about by conventional breeding, only in a much more laborious fashion, and that no part of the process involved other organisms that could cause damage or harm—so-called plant pests. So while GMOs generally involve swapping genes between different species, CRISPR can get a pass from the same level of stringent regulation as this technique can induce the kinds of genomic changes to an organism that could have arisen naturally over time.

Many applications of CRISPR-based genome editing for human gene therapy are being explored. These range from muscular skeletal diseases such as Duchenne muscular dystrophy to neurological diseases, certain types of cancer, and many genetic diseases such as sickle cell anemia and cystic fibrosis. A recent report demonstrated that a CRISPR-based approach was able to essentially cure a lethal metabolic disease known as tyrosinemia in all the mice treated. Tyrosinemia is caused by a mutation in the enzyme that breaks down the amino acid tyrosine, leading to toxic buildup of tyrosine in the bloodstream. A CRISPR-based approach effectively cured this problem by treating embryonic mice in utero. This potentially opens the door to treatments that could be applied even before a child with a particular disease is born. One key aspect of this study was that the CRISPR-treated mice were assessed for off-target effects such as genome editing of gonads and gametes, which were not specifically targeted, and none were found.

Currently, applications of CRISPR genome editing for gene therapy are being developed in such a way as to ensure that genomic changes won't be heritable—that is, they will not be able to be transferred to the next generation. Although laws and regulations vary from country to country regarding genome editing, there is a generally accepted understanding that any experimental or medical genome editing in humans should not lead to changes that could be heritable. But recent claims of gene-editing embryos in China threaten to upset the status quo.

CHAPTER 18

CRISPR Babies?

In November 2018 Chinese researcher He Jiankui of Southern University of Science and Technology in Shenzhen claimed in an interview with the Associated Press that he had successfully performed CRISPR gene editing on human embryos and that those embryos were subsequently born. This startling revelation was made after investigative reporting published in the *MIT Technology Review* suggested that such work was taking place. Subsequent presentation of some of He's results at the International Summit on Human Genome Editing in Hong Kong offered more detailed information about his claims, and prompted further consternation and concern.

According to He the twin baby girls, Lulu and Nana, had been born with differing degrees of purported gene editing. The babies were fraternal twins, not identical—meaning that they came from two separate fertilized eggs. The goal of this apparent medical (and ethical) leap was not to cure a chronic debilitating genetic disease but to give an individual resistance to HIV infection. He's experiment gained him worldwide notoriety, and dismissal from his university post. He was recently found guilty of forging ethical review forms, and sentenced to three years in prison, and fined, for his actions.

The protein CCR5 is a receptor on the surface of immune cells—in particular, those that are susceptible to infection by HIV, the CD4$^+$ (aka helper T cells). A limited number of people in the world carry a variant of this gene called *CCR5 Δ32*, which is more or less nonfunctional, and an even smaller group of individuals have two mutated copies, which essentially confers immunity to infection by HIV. He claims to have appealed to prospective parents where the father was HIV positive and offered them the possibility to spare their offspring the risk of contracting HIV.

It must be made perfectly clear that standard practice for in vitro fertilization can reduce the risk of fertilization-associated transmission of HIV to essentially zero. An HIV-positive father can provide sperm that can be used for in vitro fertilization generally without risk to the mother or child. The goal here was not to prevent the children from contracting HIV from their father but to eliminate the possibility that the children might be infected with HIV later in life. One interesting fact, however, is that the two successful embryos were not equally edited. In one, both copies of CCR5 were edited, but the other was reported to still contain one normally functional copy of CCR5 and so is still potentially susceptible to HIV infection, though there may be a somewhat decreased risk of HIV infection in people with only one mutated copy.

Further confounding factors from a medical perspective include potential increased risks from other infections that might occur in people with mutations in CCR5, including fatal outcomes from influenza virus. That being said, some researchers have suggested that He may not actually have employed the Δ32 mutation at all, and in fact created other mutations in similar regions of CCR5. The only public presentation of his data was at the conference in Hong Kong, and he has yet to offer up many scientific details for scrutiny. As these mutant variants of CCR5 do not seem to have been analyzed in detailed preliminary studies, these changes might well have unpredictable effects, raising further ethical and methodological concerns.

This information raises the question of to what extent He's actions might have been performed for medical reasons, as a proof-of-principle experiment, or simply to gain notoriety. As his results were not subjected to peer review and the conventional publication process before being revealed to the public, nobody had the opportunity to review all of the primary data before He's claims were made. Furthermore, the ethical oversight process seems to have been haphazard and lax, at best. This news was met with a general outcry regarding the potential for unintended consequences, both from the perspective of off-target effects and from possible harm caused by the mutations introduced. Significant concern over the ethical boundaries that were seemingly ignored has also been voiced.

Finally, reports that the parents of the twins were potentially led to believe they were involved in a study regarding "HIV vaccine development," and the researcher's apparent lack of experience in clinical research trials, raise further troubling questions. The outcome of proper analytical scrutiny of these claims, plus the results of a promised investigation by the Chinese Academy of Science, will hopefully be forthcoming, but for now the potential damage done to the individuals involved, and the reputation of Chinese science as a whole, can only be imagined.

It is not clear how He's actions could have been prevented, though. He apparently spoke to a few US-based experts in the field early on in his work regarding his plans to generate human babies gene edited with CRISPR. The US researchers with whom He confided claim to have vigorously tried to convince him not to move forward with his efforts, but none reported He's intentions to anyone. Although that might seem surprising, it is not clear to whom they might have spoken. One of these scientists with whom He corresponded was his former mentor, Stephen Quake, who has been completely forthright about his communications with He and whose employer, Stanford University, concluded unequivocally that Quake's actions were 100 percent appropriate. In the wake of He's shocking announcement, international bodies such as the World Health Organization are starting to discuss international guidelines and recommendations regarding this type of gene editing and possible mechanisms for reporting this type of misconduct.

Some influential scientists are suggesting a temporary moratorium on creating gene-edited humans. This raises the question of whether any application of this technology to human embryos might be acceptable. What if we're dealing with a lethal embryonic mutation, or a debilitating genetic disease? Where and how would that line be drawn? Who would draw it? What is an acceptable level of suffering that would be worth the potential risks and would also pass an agreed-upon ethical standard?

Biomedical ethics is a relatively mature field that has dealt with complex questions like this before. However, it is clear that in He's case the proper procedures were not followed. And as he was not in any way sparing these children from any certain medical issue, the overall rationale for his actions was not sufficient.

Another outcome of rogue scientists working outside accepted norms is that their actions might jeopardize the public's faith in more conventional and beneficial activities in similar areas. Thus, a moratorium or ban on gene editing might be too reactionary and ultimately restrictive, since CRISPR-based therapies could have real value to people suffering the effects of debilitating genetic diseases.

Where do we go from here? Should He's findings be published in a conventional scientific paper so that other researchers in the field have a chance to determine exactly what was done? Should this simply be treated as any other case of scientific misconduct? If so, this generally means an internal investigation by the institution that employed the researcher, and possibly other entities like funding bodies and scientific journals. These types of misconduct investigations sometimes also lead to legal action, though they do not generally involve international bodies. Ultimately, it would be up to the legal system of a specific country, as well as funding bodies and research institutions, to create and enforce relevant mechanisms that might spell out exactly what is acceptable, and the consequences if scientists act contrary to these guidelines, rules, regulations, and laws.

He's actions appear indefensible, both scientifically and ethically. How this was able to occur, and what might stop the next person who wants to grab the spotlight with fantastic scientific claims, is not clear. However, this is not the only example of genetic modification that leaves the public skeptical, and unfortunately it won't be the last. Agriculture is another area where there is already a long-standing debate over genome editing. The topic of GMOs has long been contentious and capable of generating extremely passionate debate. Similar to the question of CRISPR babies, GMOs raise important technical and ethical questions; however, unlike He's actions, scientific consensus seems clearly on the side of genetically modifying organisms to improve our abilities to thrive as a species in the face of an uncertain future with a rapidly changing climate, and an increasing global population.

CHAPTER 19

GMOs–Genetically Modified Organisms

A genetically modified organism (GMO) is commonly taken to refer to a plant or animal that has had its genome modified through laboratory-mediated techniques to bring about a particular change. Furthermore, if that change is heritable, it can be passed down to subsequent generations.

Most organisms targeted for genetic modification have two copies of each gene. However, if you simply want to add a gene from one species to another—for example, one responsible for producing a compound that prevents insects from eating a crop—it might not be absolutely necessary to engineer two copies into the genome. Once altered organisms start breeding, however, having both gene copies altered permits continuous reproduction of the modified strain, without risk of what is referred to by geneticists as "loss of heterozygosity." This means going from having a single copy of the altered gene to none.

If two heterozygous individuals mate, each containing one naturally occurring and one altered version of a gene, or with only one copy of a newly added exogenous transgene (for example, from another species), then 25 percent of the offspring will not carry the gene, 25 percent will have two copies, and 50 percent will end up just like the parents. This is what is referred to as Mendelian inheritance, named after the father of modern genetics, Gregor Mendel (see p. 127). By having homozygous individuals with two copies of the transgene, these inheritance variations are obviated as all progeny will be the same as the parents, and the line will "breed true."

Because bacteria don't have paired maternal and paternal chromosomes, they generally only have a single copy of each gene, unless for some reason that gene has been duplicated. So for bacteria only one copy of the transgene should be required. It should be noted, however, that bacterial genomes can be very easily changed when the presence of plasmids (small circular pieces of DNA that can exist independently of the larger chromosomal DNA) are taken into consideration. Plasmid DNA is easily added or lost from bacteria and can be shuttled between individual bacteria in a population in a process referred to as "horizontal gene transfer." Understanding the differences when discussing bacterial GMOs is critical. From bacteria that convert milk to yogurt to those that have been engineered to produce human insulin, the applications for genetically modified bacteria are huge, and the ease of propagation makes them well suited to industrial-scale production.

GENERATING GMOS

The methods of creating GMOs vary widely, and innovation in this area is moving rapidly. The techniques employed can depend greatly upon the specific organism being altered, as well as the type of change being made, whether it be genetic deletion, alteration, or addition. As described above, classical genetic engineering has usually focused on the challenge of getting foreign DNA into a cell, and then relied on very low-probability insertion or replacement events. The efficiency and precision of genetic modification has recently taken a huge step forward with the development of CRISPR techniques (see p. 201).

The application of CRISPR to producing GMOs has implications that are only now being considered. However, human activity aimed at making heritable alterations to the genome of another organism is not new. Humans have long sought to alter the genomes of other species. In fact, humans have engaged in genetic engineering for thousands of years. While today the

CRISPR "revolution" is on everybody's lips, research scientists have been developing a host of techniques for deleting genes or changing their sequence for many years. Indeed, long before there were geneticists or molecular biologists, humans employed the tools of selective cultivation, agriculture, and domestication to modify the characteristics of the living world around us through manipulating genetics and heritable characteristics. Think of the horse breeder who mates his fastest stallion and mare to maintain a successful line of race horses.

When we walk around a supermarket, or go to a farm, we are visiting museums of applied genetics. In fact, the primary difference between the techniques employed in labs today, which we call "genetic engineering" and create GMOs, and historical methods are that the former are more efficient, predictable, and precise. Selectively transferring a gene from one organism into another is generally not feasible when employing traditional agricultural methods, but selecting individuals with advantageous characteristics relative to a specific time and place has occurred in nature as long as there has been life on earth. This is the basic definition of evolution. Humans have been directing this process for thousands of years—first in the fields, and now in the lab. So why the sudden panic?

Those giant sweet ears of corn slathered in butter and salt that we munch every summer didn't really exist until humans came along to selectively propagate the most desirable specimens available of the wild precursors to maize. Given all the concern over GMOs in our food system, it probably makes sense to stop and consider that all corn (even before any lab-directed modifications) can be looked at as an organism that was genetically modified through human activity. Humans living in what is now Mexico discovered that teosinte, a type of wild grass, contained a small pod with a few rows of edible seeds. Through thousands of years of selecting seeds from plants with the most desirable characteristics, corn emerged as a distinct species with a greater number of larger, sweeter kernels that are easier to process and consume.

Many organisms that we know today don't have an exact wild counterpart. Isn't it amazing that simply through selective breeding wolves have become Pomeranians, Chihuahuas, and Great Danes? Through agriculture, domestication, and cultivation, human activity has shaped the world we live in. The tomato that we associate so firmly with Italian food was only introduced to Europeans less than five hundred years ago and started out as a tiny fruit cultivated for thousands of years in South America.

Even the origins of the modern understanding of genetics comes from studies of cultivated plants. Gregor Mendel was a monk studying what made some pea plants tall or short when he devised the basic understanding of genetics that is still applied today. The peas that Mendel used in his experiments have been cultivated by humans for thousands of years and have a significantly different genetic basis from their closest wild relative.

When Darwin was comparing the beaks of finches in the Galapagos and correlating the size and shape to the specific diets of each type of bird, his understanding of "the origin of species" through evolution was fixed upon natural selection—that is, the reproductive success of individuals with genetic variants (mutations) that make them better than others at exploiting a specific environment or resource. Is there really a difference between a scientist making "Frankenfood" in a lab—for example, a drought-resistant strain of corn that can be grown in desert climates—and a woodpecker evolving a long, thin beak to pull insects out of trees better? What are the true risks of GMOs, and is the near hysteria surrounding the generation, propagation, usage, and consumption of GMOs warranted?

We also should be mindful not to fall into the opposite extreme and take it for granted that GMOs will be the panacea for all disease and end world hunger, overpopulation, and concerns over access to fresh water. Without careful attention to detail, there are risks associated with GMOs, though they're not the same as what people tend to worry about. One major valid concern is proper isolation of GMOs, particularly in agriculture or animal husbandry. This is an issue that laboratory scientists worry about all the time; only we refer to it colloquially as "cross-contamination." If you are growing two types of cells in the lab, or culturing different strains of bacteria, you do everything you can to ensure that the different samples don't get mixed. You use fresh lab supplies and might conduct studies in specially designed isolation and containment areas.

Imagine that you need to prepare both hot sauce and baby food. You would want to make very sure that they don't mix. You might not want to make them in the same place at the same time. You certainly wouldn't want to stir them with the same spoon! So if a farmer is planting GMO corn in a field adjacent to non-GMO corn, care needs to be taken to ensure that cross-pollination doesn't occur. Failure to do so could result in significant financial consequences if one batch of corn was meant for a special purpose, such as sale in a country where GMOs are banned. As long as proper procedures are followed, the risk of intermixing GMOs with non-GMOs can be controlled to a large degree.

That being said, it can be a lot harder to keep an animal GMO from reproducing with domesticated or wild relatives. The AquAdvantage salmon has been genetically engineered with an increased rate of growth so that it reaches a mature commercial weight in half the time as the naturally occurring species. Allowing this type of product to be developed, produced, marketed, and consumed certainly should not be taken lightly, and it wasn't. In fact, it took approximately ten years for the US Food and Drug Administration to approve AquAdvantage salmon for the retail market. And even after all that, AquAdvantage salmon is still not for sale at your local fish counter. Decisions on how to regulate GMOs, and how to label these products, have been developing. However, with the recent passage of the National Bioengineered Food Disclosure Standard, which regulated the labeling of GMO foods, the US Department of Agriculture (USDA) is now poised to develop and implement regulatory frameworks for these types of products.

In an attempt to ensure that mixing with wild populations does not occur, the fish are raised in tanks on land, and only sterile female salmon are cultivated. It would seem that every precaution has been taken to ensure that these fish will not escape into the sea and mate with wild salmon, as this could permanently alter the population worldwide. Of course, the similarities between this case and certain plot points of *Jurassic Park* are hard to miss.

OFF-TARGET EFFECTS IN GMOS

Several other concerns regarding GMOs are more manageable through scientific methodology and analysis. In some cases, to introduce, alter, or delete a specific gene, other genetic material apart from the gene itself is required as part of the overall "construct." Often this includes sequences that regulate expression, and contain multiple different sites for restriction enzymes to cut the DNA to provide insertion sites for foreign genes, antibiotic resistance genes that allow for selection of organisms that have taken up the construct, and markers, such as a gene for a fluorescent protein that allows transgenic cells to be visualized. Addition of these flanking sequences alongside the transgene can have unintended consequences.

This is like if you placed some meat into a frying pan, not noticing that that little pad meant to absorb liquid in the package is still attached, or cooked a turkey or chicken with giblets still in the bag inside the bird. Something else went along for the ride. Sometimes it is unavoidable that sequences of DNA adjacent to the gene of interest are altered, added, or removed. This could conceivably have unintended consequences, or "off-target effects." These can include insertion at sites other than what is intended, which can potentially disrupt a critical gene, or introduction of sequences other than the specific transgene, such as the gene encoding antibiotic resistance or the fluorescent protein.

Because of such considerations, great care needs to be taken with any type of gene therapy, which generally aims to alter a patient's genome in order to cure a genetic disease. In particular, changes to the human genome that might be inherited are being looked at with extra scrutiny. If we are OK placing genes into humans to save their lives, how is it that engineering an apple that doesn't bruise so easily has become such a bone of contention?

Of course, the interpretation of potential risks of GMOs can vary, and the United States, the European Union, and several other nations, are taking different approaches to regulating for GMO development, production, and sale. While the United States is seemingly adopting a cautious but progressive perspective regarding GMOs, there are still many conflicts and compromises regarding regulation, labeling, and product marketing. However, in Europe the approach to GMOs has been very different. One assessment conducted in 2017 revealed that the amount of land being employed for growing GMOs in the United States was over five hundred times higher than in the European Union. This is a staggering difference, especially considering that the contiguous United States is less than twice as big as the European Union by area.

Looking at the evidence from a scientific point of view, the apparently negative feelings toward GMOs can be quite hard to understand. Humans would simply not have become what we are today without all the genetically manipulated organisms that have evolved alongside us. To put it bluntly, consuming GMOs poses virtually no real danger to humans.

Plant genetics are very well understood, and it is extremely easy to perform genetic analyses on crops, as well as conduct assessments for unwanted effects of genetic manipulation. So how did GMOs get such a bad reputation? Why are countries banning GMOs? Why are companies being forced to label products that contain GMOs? Why are some companies choosing to specifically highlight if products are "GMO free"? If GMOs can potentially reduce pesticide use, and water requirements, and make it easier to grow more food for people, why are so many environmentalists against them?

One answer might reflect the way that many people first were introduced to GMOs, at least in the media and marketplace. Roundup is an herbicide, a weed killer, based on the chemical glyphosate. If you are trying to grow plants for food, why waste water and fertilizer on weeds? It makes sense to want to get rid of them. However, indiscriminately spraying herbicide on your farm will kill your crops, as well as the weeds. Therefore, plants have been genetically engineered that are resistant to Roundup. These "Roundup Ready" crops can be sprayed with the herbicide, which will then selectively kill only the weeds. However, an innovation that ends up permitting more liberal herbicide spraying is not an environmentally friendly outcome. One reason is that herbicides sprayed on crops can later end up in unintended places, like rivers, streams, and people. And by killing more plants, even those designated as weeds, you decrease biodiversity and indirectly affect pollinators like bees and butterflies. This is especially troubling given the significant concerns about the potential for human toxicity from glyphosate. To make matters worse, with all this spraying, it seems the weeds are evolving resistance to Roundup, too. The story of Roundup Ready crops is not the best advertisement for GMOs.

That being said, from drought-resistant crops to plants that inherently produce compounds that repel insects, overall the scientific consensus is completely clear: GMOs are a net benefit for humanity, and banning and fearing them could be considered akin to knee-jerk denial of human-caused climate change, the critical benefits of vaccines, or the theory of evolution. As humans, we unavoidably alter the world we live in. It is clear that there have been some missteps in the application of GMOs, and there are certainly some risks. On balance, though, banning them out of abject fear will ultimately do more harm than good. The irony here, of course, is that use of GMOs pits the environmentalists who, when it comes to climate change, argue for science against corporations but, when it comes to GMOs, corporations are on the side of science.

There are countless examples of the benefits of GMO foods. One gaining significant attention is Bt, short for *Bacillus thuringiensis*, a naturally occurring bacteria that produces a chemical that is toxic to many of the insects that devour crops. So-called Bt plants are genetically engineered to produce this chemical, which is actually a component of some pesticides, and thus require less spraying. Importantly, the insecticidal compound from Bt has been safely employed in agriculture for decades and has been shown to be nontoxic when consumed by humans. Another important example of a critically important GMO is Golden Rice, which is genetically engineered to produce beta-carotene, which is responsible for the "golden" color and serves as a precursor to vitamin A production. In many developing countries vitamin A deficiency can cause blindness and even death. Thus, Golden Rice could have a very significant impact in these areas. However, the skepticism and fear regarding GMOs has greatly hindered implementation in many places where they could make a huge impact, ranging from developing nations such as India to the developed world such as the European Union.

Where does the blame lie for the animosity toward GMOs? Why would anyone want to stand in the way of progress in such a way as to actively withhold the benefits of GMOs— drought-resistant and insect-repelling crops, and so on? It isn't clear yet to what extent misinformation, misplaced good intentions, or other ulterior motives are to blame. However, considering that famine and malnutrition are such huge problems worldwide and are continuously getting worse, the general lack of enthusiasm toward GMO adoption into worldwide food production can be viewed as a victory in the war against science and reason.

Of course, as seen with Roundup Ready crops, not all the players in GMO production and dissemination will turn out to display holistically beneficial motives. Furthermore, the power of corporations to manipulate the public is another interesting wrinkle in this ongoing saga. This does not mean we should throw the baby out with the bathwater and cave to anti-GMO hysteria that grossly overemphasizes the potential negatives and ignores the positives.

If people want to know they are consuming GMOs, for whatever reason, it is hard to argue that the information should be hidden from them. However, while labeling GMOs is being mandated across the United States by 2020, it's most likely this will take the form of a smiling sun declaring the much less scary euphemism "bioengineered." Deliberately ambiguous, confusing, or even irrelevant product labeling can blur the lines between public awareness and marketing. Labeling a product "non-GMO" seems to be an unregulated marketing stunt with zero basis in science or governmental regulation. This is about as informative as labeling a bottle of water "gluten free." Interestingly, however, since Vermont started labeling GMO-containing foods a few years ago, it has been suggested that attitudes of consumers in that state may be becoming less negative toward GMOs. Whether anti-GMO attitudes are winning in the USA remains to be seen. At the end of the day, when environmentalists are not looking at the science, and corporations are drivers for the public good, things are starting to look a bit topsy-turvy.

If GMOs are neither a panacea that will solve all our problems nor a threat to all life on earth, what might the future of GMOs look like? How might we employ the tools of genomics and genetic engineering to benefit humanity and ease suffering? One key perspective that is emerging involves comparing domesticated cultivated organisms and their wild relatives, and the potential application of this information to better develop crops to feed our species.

CHAPTER 20

The Future of Food?

The process of the domestication of wild species through selective cultivation and breeding of those with particularly attractive and useful traits ultimately leads to genomic changes that are passed down to the next generation. This process of modifying plants to suit our needs began thousands of years ago with many of the crops that continue to support our species. Although these activities might not have been well organized, informed, or deliberate, selecting superior individual plants according to certain criteria led, over time, to the emergence of many of the crops we grow today.

In some cases, random wildly occurring mutations made plants that were previously inappropriate for widespread human consumption into suitable, nutritious foods. The predecessors to the almonds that humans grow and enjoy are extremely bitter and produce a potentially toxic substance known as amygdalin that is actually broken down into cyanide during digestion. A naturally occurring dominant mutation resulted in almonds that are safe to eat and don't make significant amounts of amygdalin, and these are what humans have been cultivating for thousands of years.

Generally speaking, the transition from wild plant to cultivated crop can take a standard trajectory. Known as "domestication syndrome," similar traits were selected for across many individual species that made a plant easier to grow, harvest, process, and consume more efficiently. Characteristics such as size, shape, and taste have been improved in many food crops through this type of gradual selection and domestication. More recent intensive agricultural practices have further focused on developing particular strains that are suited to specific environments, or that display certain traits such as uniformity or shelf life.

This type of focused selection decreases genetic diversity, and can increase the emergence of problematic mutations, such as reliance on very specific conditions for growth. It can be difficult to adapt domesticated species to changes in environmental conditions. Targeted genomic changes can domesticate wild species much more rapidly than might otherwise occur, and by selective breeding in a more directed fashion, crops particularly suited to specific environments could be developed. As seen above in the case of the ground-cherry, the advent of tools for genomic manipulation and genetic engineering like CRISPR has provided the means for rapid and directed development of domesticated species suitable for cultivation and distribution. However, determining the specific genes and variants to focus on requires significant understanding of not only the processes that govern domestication, such as efficient cultivation and resistance to pests, but also the mechanisms underlying the ability of wild species to adapt to changing environmental conditions. By analyzing the wild relatives of the domesticated crops we already rely on, it might be possible to make improvements in species that are currently cultivated.

The genetic diversity within wild species provides the basis for successful spread into areas that differ in nutrient availability, temperature, humidity, and elevation, as well as adaptation over time—for example, during periods of rapid climate change. Domesticated species, whether developed over eons through gradual selective pressure, or overnight in a lab, tend to lack the genetic diversity to facilitate this type of flexibility, making them much more sensitive to changing conditions and ultimately limiting them to thriving in specific environments. Wild relatives of domesticated species can give insight into particular variants that might potentially be useful for increasing diversity, facilitating expansion into new areas, and maintaining continued cultivation when changes occur in temperature, water availability, the presence of pests, and other biotic and abiotic factors.

One way of looking at this issue is to consider the process of baking bread. There are many variables that determine whether you successfully bake a loaf of bread: the protein content of the flour, the age of the yeast, temperature (both for rising and baking), the type of water used (purified, or hard or soft water from the tap), whether you are using finely ground salt or coarse crystals, and so on. All these subtle factors work together to determine how well your bread will turn out. A novice baker relying on a recipe would not be able to adapt if any variables changed. You might be able to bake a nice loaf at home in your kitchen, but if you went to another location with different ingredients and conditions, your bread probably would not turn out the same.

The way bread baked by a novice reacts to variations in the ingredients and conditions is like the domesticated crop that lacks the genetic diversity to handle alterations in the environment. The expert professional baker knows that if it is particularly humid, you add a touch less water. When the room is colder, you have to let the bread rise a little longer. This adaptability is like the wild relative of the domesticated crop that still possesses the necessary genetic diversity to deal with climate change and thrive in a variety of environments.

To help the novice baker, you might write further instructions into a recipe. For example, you might list two rising times, one for warm room temperatures and one for cold. This is comparable to using genome editing to introduce one new trait that might permit growth under one particular form of stress—say, higher temperatures or drier soil. Adding one additional alternative might help if only a single variable changes in a particular way. However, this doesn't solve the larger problem of genetic diversity, and although one alteration or variable might work in the short term, it will not create overall flexibility. In fact, it might end up locking the modified plant into thriving only under one specific growth condition, and ultimately further reduce genetic diversity. What is the better strategy then? How can we reverse engineer wild plants to help make genetically uniform crops more diverse overall? Is there a middle ground between a GMO specifically bred for high productivity or some other specific trait, and the flexibility and diversity found in wild crop relatives? One solution may be cross-breeding the two and taking care to maintain the trait of interest. Another could be introducing key wild gene variants into the domesticated crop in a more targeted fashion.

Collecting wild crop relatives from a variety of environments can facilitate engineering new crops that will succeed as climate and conditions change. Considering that many of the crops most critical to human survival were first domesticated thousands of years ago in areas that are very different from where they are currently grown, understanding the mechanisms behind adaptation of wild species to changing environmental conditions, as well as facilitating cultivation, is critical, especially as the direct and indirect effects of climate change become more widespread. The time to act is now, while the wild populations that are the reservoirs of genetic diversity are under threat. Thankfully, repositories of wild seeds are being maintained and tools for genomic analyses of wild and domesticated species have recently been developed. In addition to classical breeding techniques, the ability to employ techniques like CRISPR to perform focused genetic manipulation of livestock and key food crops is within our grasp. As the human population grows and the environment becomes increasingly fragile and unpredictable, the question that remains is how to perform combined genomic and trait analyses comparing domesticated species with their wild relatives, and how to apply this knowledge to increase food security.

Douglas Cook is a researcher from UC Davis who is doing exactly this. His focus is primarily chickpeas (aka garbanzo beans), which are a significant food source for much of the world. Chickpeas were first domesticated in the Fertile Crescent in ancient Mesopotamia approximately ten thousand years ago. Wild relatives of chickpeas still grow in parts of Turkey with a high degree of genetic diversity that supports adaptation to varied environmental conditions. Cook's group has collected specimens of wild relatives of chickpeas in Turkey, analyzed them for the genetic ability to adapt to different types of environments, and compared these to the domesticated crop being grown in farms around the world. Working together with ecologists, they have collected wild relatives of chickpeas from a variety of different environments and analyzed them to try to understand how genetic diversity has permitted the plants to succeed in response to an array of stresses.

The samples were analyzed through a technique known as genotyping by sequencing, which compares similar regions across the genomes of multiple related species. You can think of the process like comparing multiple translations of the same classical text, such as the original *Odyssey* in Greek, to identify not the names, places, and plot elements that are the same across all different translations but the literary flourishes that make the different translations unique.

Exome sequencing is not preferable for this type of work because it biases the results by focusing only on protein-coding regions. Relevant variants are in enhancers or repressors, or other parts of the genome that regulate expression level, and would be missed in exome sequencing. However, a study using genotyping by sequencing is primarily employed to test the hypothesis that variable genomic regions are involved in permitting adaptation to diverse environments, and to that specific end only a small proportion of the total genome needs to be analyzed. In this case, only about 5 percent of the total genome was analyzed, but ultimately whole genome sequencing (WGS) might be required to move the work forward—for example, to identify specific gene variants responsible for particular traits.

Currently Cook and his team are attempting to understand the mechanisms behind the flexibility of plants to adapt holistically to different environments, rather than developing specific GMOs that will introduce a single trait into a crop. The ultimate goal for this work would be to create a paradigm of comparative analysis, identification of target genes and regulatory sequences, and then genomic manipulation in order to ensure food security for the human species moving forward.

Heritable genetic variation is the fuel for evolution and permits individual members of a species to succeed when faced with various selective pressures. One interesting discovery of this study was that genetic diversity in wild relatives, such as the presence of SNPs among different plants, was much greater in the wild relatives of chickpeas compared to the domesticated crop. They further found that many of these genomic variations actually encode alternative protein sequences, and potentially have direct influence on specific traits that are key to survival in different areas and under various conditions.

Specifically, they found that wild chickpeas from drier areas had adapted by increasing their ability to conserve water. Wild plants were also more resistant to a beetle that is a major problem for domesticated chickpeas. Genomic analysis of the wild relatives of chickpeas and comparison with the domesticated crop have provided specific insights into how to potentially develop plants that are better suited to handle environmental stresses. They found that hybridization between different wild chickpea relatives can occur, and this has implications for potentially increasing genetic diversity in the domesticated crop. As proof of principle, they then crossbred wild plants with the domesticated crops and found that the progeny were more resistant to heat stress, a key concern as we face increasing global temperatures. These studies have paved the way toward our ability to develop crops that are better suited to a future where we need to feed more people more efficiently, as well as one where the environmental conditions are not well suited to the cultivated crops on which we rely.

Much more work is required to make Cook's work a widespread reality. But the benefits of applying genomics to related wild and domesticated species are clear. Although less critical to our survival as a species, many of these same tools for genomic analysis, and potentially genetic engineering, are currently being applied to the domesticated animals with which we live—our pets.

CHAPTER 21

Pets–Ancestry, Health, and Cloning

W e love our pets and often look for ways to apply new technologies to improve their lives. Genome-based studies are beginning to be extended beyond manipulating crop and farm animals, and investigating the ancestry and health of humans, and have recently begun to be employed in an attempt to better understand our favorite nonhuman companions. From trying to understand the origins of domesticated animals, to efforts at cloning beloved pets after they have died, to applying personal genetic testing to pets in a veterinary health context, we are rapidly applying the tools of genomics, in more or less reasonable ways, to the animals with whom we share our lives.

Recently, by applying the tools of genomics, fascinating insights into the origins of domesticated dogs in America have come to light. Through a combination of classical anthropological research and modern genomic analyses, a very interesting picture is starting to emerge. As most people know, modern domesticated dogs originated from wild wolves. Similar to the way that modern sweet corn now bears little resemblance to the wild Mesoamerican grass teosinte, dogs were domesticated from wolves through the actions of humans. The relationship with humans, over time, has changed their bodies and their personalities. Some of the differences between wolves and dogs are more than skin deep. It was recently revealed that muscular changes evolved in dogs that allows them to look at us with those heart-melting "puppy-dog eyes."

Although *Canis lupus familiaris* technically represents a distinct subspecies of their wild forebears, wolves and dogs can still mate and produce fertile offspring. This is quite significant because, in many cases where crossbreeding between species is possible, it does not result in offspring with the capacity to breed, such as the case of mules, which are the results of crossbreeding between horses and donkeys and are almost always infertile. Furthermore, genomic plasticity has permitted humans to employ selective breeding to unlock the huge heterogeneity observed in dogs, as Great Danes and Chihuahuas remain the same species (albeit different breeds) despite the dramatic difference in their sizes and other physical characteristics. Thanks in no small part to using genomic technologies on archeological specimens, a picture has begun to emerge of where and when wolves were first domesticated and how dogs became the companions they are today.

In reality, the story of dogs in America is actually quite complicated, and also potentially a bit upsetting. A recent large-scale analysis of DNA sequences derived from ancient and existing dogs paints a complex picture of when and where dogs were domesticated. It seems that dogs may have been first domesticated from Siberian wolves somewhere around fifteen thousand years ago. While this roughly corresponds with the start of human migration into the Americas, archeological sites containing dog remains don't begin to appear in the New World until about five thousand years later. It is unclear whether, like Odysseus did when he left his dog, Argos, for Troy, the first humans to enter the Americas left their dogs behind in Siberia. In addition to the archeological record, the accounts of the first Europeans described native dogs, and pre-Colombian Mesoamerican art contains unmistakable images of domesticated dogs. Clearly, there were dogs present in the Americas nearly ten thousand years before Europeans arrived, and they were domesticated before migration, most likely from Siberian wolves, rather than from American wolves. These first-wave American dogs seem to have spread across the continent down into Mesoamerican cultures bridging North and South America.

Interestingly, sequence analyses comparing DNA from modern dogs with samples extracted from the archeological remains of ancient domesticated American dogs revealed essentially no direct connection. Overall, the two groups shared extremely little genetic information, only a few percent of the total analyzed. This would suggest that dogs in present-day America did not in any way originate from the existing pre-Colombian native populations and that the indigenous dogs did not significantly interbreed with those that came over from Europe. This recent comprehensive analysis may not be completely consistent with some earlier more limited evidence suggesting that some American dog breeds have retained significant genetic links to ancient pre-Colombian ancestors. These different studies employed different sample types, sample sizes, and analytical methods, so the complete picture might still be emerging. At the end of the day, though, it seems domesticated dogs in America are descended from European dogs, which began to arrive as early as their European masters, including on the *Mayflower*.

If the domesticated dogs we currently have as pets in America essentially don't share any ancestry with the dogs that were here before European contact, what happened to the dogs that had been domesticated by Native peoples? If the Native Americans had dogs that originated from wolves domesticated before European migration, and the Europeans brought over their own canine companions, which are the direct forebears of our own dogs, where did those native dogs go? What happened to them? Were they exterminated? Were they slaughtered along with the Native American peoples, or decimated like the buffalo? Did they succumb to disease carried over from European dogs, similar to the widespread epidemic of smallpox that killed off many Native Americans? This critical piece of the puzzle remains unsolved. It is possible that some isolated breeds in America still carry more direct genetic ties to the original pre-Columbian populations, and maybe over time more thorough analyses will resolve these open questions.

Another largely open question regarding domesticated dogs is how human activity could functionally change both the genes and behaviors of wild animals, and how the two might be linked. Although foxes might seem an unlikely place to find insights into dog domestication, as the two species diverged approximately ten million years ago, they have actually retained roughly similar genomic architectures. Importantly, simply raising foxes in captivity will not result in domestication. Rather, selective breeding for approximately ten generations is required to generate tame, truly domesticated foxes. Researchers have recently performed genomic comparisons among different groups of foxes and have identified thirty specific regions containing particular SNPs that correlate with domestication. These regions might contain gene variants associated with domesticated behaviors. Some recent work has questioned aspects of the classical perspective regarding fox domestication and further analyses are certainly needed, especially to test for any connection with the genetics of dog domestication, but this is a very interesting start.

Cat domestication, however, is a totally different story; in fact, some studies seem to question whether humans ever really properly domesticated cats. The available genetic evidence seems to suggest that small wildcats actually chose to begin their association with humans, rather than being subject to active breeding to develop the traits associated with domestication. As anyone who has ever lived with a cat can attest, this explanation seems to fit well with the independent and seemingly feral personalities of many pet cats that appear to barely tolerate our presence.

VETERINARY TESTING

Other genomic analyses aimed at uncovering aspects of canine biology are significantly more focused, and in some cases potentially questionable. Genetic screening is used to assist breeders in producing purebred puppies and to assess medical conditions. It is also being used to help owners of mixed-breed dogs trace genetic ancestry and understand the parentage of their beloved mutts. Generally speaking, most of the techniques focusing on veterinary issues are taking a candidate gene approach rather than employing large-scale genome-wide association studies (GWAS) in an attempt to look for SNP patterns correlated with a particular disease risk. Direct-to-consumer genetic testing kits for dogs are being employed in a similar fashion to the 23andMe Genetic Health Risk tests, albeit with many more limitations and potential caveats.

When relative risk analyses are performed, it seems as though, in many cases, the candidate genes are identified as members of pathways that might be involved in a disease process—for example, clotting factors or enzymes involved in cell proliferation, which could be involved in hemophilia or cancer, respectively. In addition, the evidence supporting a role in disease risk for a particular variant is generally correlative at best, not causative. However, each and every polymorphism will not definitely increase risk significantly, even in genes that have been directly linked to specific diseases. In human genetics this type of candidate gene approach, in the absence of confirmatory data, is not accepted as strong evidence that a specific gene variant is truly a significant risk factor for a particular disease. There are too many variables and multiple small studies claiming to have identified disease-associated mutations that have not held up under further scrutiny. Even assuming the scientific basis for the test is accurate and the interpretation of the results is correct, it can be hard for pet owners to focus on the fact that what is being tested for are generally polymorphic variants that *might* be associated with a somewhat increased risk of developing a complex disease. Basically, these screens are not necessarily diagnostic tests for disease-causing dominant or recessive mutations.

There are many other potential issues with using the results of these tests to make medical decisions for our pets. The basic science studies that underlie selecting candidate genes may not have been performed with the same level of rigor as is done in the case of human biomedical research. Veterinarians are not trained to perform the type of genetic interpretation and consultation required. Interestingly, although these genetic screening kits may have been initially marketed through veterinary clinics, they are now more often sold directly to the consumer. Also, while the FDA regulates human genetic testing for disease-risk screening, this is not the case with direct-to-consumer tests aimed at pets. However, it seems that industry groups are beginning to come together to enforce standards, and to pool databases. Unfortunately, though, it is generally the same veterinary medical companies that produce these screening kits that stand to profit from resulting therapeutic interventions, which raises significant concerns about conflicts of interest.

CLONING YOUR PET

Another application of genomic technologies to pets is turning into big business among a certain group of bereaved pet owners. If you provide fresh tissue within about five days of a dog's passing that contains intact DNA—along with about $50,000—veterinary scientists can clone your dog. The cloning process takes about five months and occurs with a success rate of about one in twelve, meaning that for every twelve tries, one viable clone will be born. The procedure is essentially the same as that employed to produce Dolly the sheep over twenty years ago. However, it reportedly took 277 attempts before Dolly was successfully born.

Initially, dogs were difficult to clone. The first cloned dog was produced in 2005 at Seoul National University (SNU) and was an Afghan Hound named Snuppy, a contraction of *SNU* and *puppy*. It took approximately three years of work and over a thousand donor eggs to generate Snuppy. The way the cloning works is that an undamaged nucleus is removed from a cell obtained from the animal to be cloned and then inserted into an egg from another animal, from which the nucleus has been removed and discarded. This new hybrid cell is then implanted into a third animal treated with hormones to induce viable pregnancy, and if the cloned embryo successfully implants into the uterus and gestates, a clone can be born.

It is important to note that clones created in this manner are not completely identical to the original animal. There are aspects of genetics that are not encoded within chromosomal DNA, such as the genes that are present in mitochondrial DNA (see p. 275), which will not be carried over during the nuclear transfer step. Furthermore, chemical changes to chromosomes that can regulate gene expression, referred to broadly as epigenetics, change over time and won't necessarily be properly regulated when normal fertilization and development are circumvented this way. In many ways, the clone is not the same as the dog that donated the cell nucleus. These differences can be readily apparent from the personality of the cloned animal and from physical attributes such as coat marking patterns, which are controlled in part through epigenetic mechanisms. However, for people greatly distraught over their lost canine companions, and who have significant disposable income available, pet cloning can bring some level of happiness and relief from grief.

Although the application of genomic technologies to the world of human pets and domesticated animals has great potential to provide valuable insights into the health of animals, there are certainly ways in which unscrupulous actors or sloppy science could have significant negative impacts. One horrible story that should give us all pause involved a dog that was suffering from a painful but potentially treatable spinal condition. The dog's owners used a direct-to-consumer genetic test, which showed that the dog carried a mutation linked to progressive neurodegenerative disease. The owners chose to euthanize the dog. However, what the owners didn't understand was that only a very small percentage of dogs that carry this mutation will actually develop that disease, at a rate of something like one in a hundred. Thus, it is very possible the dog did not actually have that disease and did not need to be put down, but limited understanding of the true meaning of the genetic test led the owners to make a decision that could not be undone.

The limited scientific basis of many genetic tests sold for pets, the inability of most people to understand results without the help of expert consultation, and the lack of regulation and oversight seem to be generating a potentially tragic mix of conditions that will likely lead to more unfortunate stories like this. Given that Americans spent approximately $50 billion dollars on pet care in 2018, it seems that marketing genomic technologies to pet owners is a potential growth industry. Without government intervention or significant market-driven demand for better standards, it is hard to think that profit won't be the main driver of development in this area. That being said, a great deal of excellent science is emerging, and analyses of pet domestication are proving to be fascinating.

Some of these findings have potential impact beyond domestic animals. How are advances in genome science playing out with undomesticated wild organisms? How do genomic technologies help us with environmental conservation of living plants and animals? Is there a potential role for these tools to assist us with eradicating pests and disease-causing organisms that pose significant risk to humanity and our interests? If so, is that an ethical or wise use of the power of genomics?

CHAPTER 22

Conservation and Eradication

I t is an inescapable reality that humans shape the environment. From global climate change to pollution, plastic in the oceans to fertilizer runoff in streams, our species has a pretty lousy track record in environmental stewardship. Nonetheless, there are many individuals who are devoted to limiting the environmental impact of humans and possibly improving conditions here on earth through initiatives to help clean up the environment and reduce further damage. One major focus of the environmental movement are conservation efforts aimed at maintaining the wild plants and animals in the world around us in the face of mass extinction. There are a number of rationales for this idea, including that plants and animals are required for our continuous existence. More altruistically, many believe that we simply have an ethical duty to support being part of the solution, rather than contributing to the problem.

Conservation in this specific context primarily refers to maintaining the living world around us, although issues such as erosion, temperature, sea level, and weather events certainly are part of the discussion. On a global macroscale, and also within individual ecosystems, the biological and the nonliving abiotic components of our world are interrelated in so many direct and indirect ways. The pertinent question in the context of this book is addressing how genomic technologies might assist with these conservation efforts. On a related note, how can the tools of genomics be harnessed to modify the living world in ways that are beneficial to our species? Might that involve eradicating harmful species? Manipulating wild populations of organisms in this way through genomic means could certainly be controversial. In each case, conservation and eradication, what do the potential unintended consequences look like on a global scale, and what role does ethics have to play in deploying genomic technologies into the wild?

CONSERVATION EFFORTS

As the global climate changes, temperatures increase, glaciers melt, sea levels rise, certain coastal areas become wetter, and landlocked regions become drier and more arid, the environmental pressures on wild species will grow. Genomic tools can be employed to understand how different species, populations, and individuals may be able to cope with these changes, while others might not. On a more macro level, genomic efforts can take the form of surveillance over time of the different groups that are present in an ecosystem, or they can be applied to more focused and directed analyses of the specific genes and mutations that might permit survival as conditions change. Given our absolute reliance upon other species for our continued existence, the power of genomics to assist with conservation efforts cannot be overlooked. If these tools could help us ensure that there are enough oxygen-producing green plants, pollinators to assist our agricultural efforts, and an ongoing supply of edible plants and animals, ensuring the development and application of effective and efficient genomic technologies must be a priority in conservation.

But a number of issues arise that increase the complexity of applying genomic technologies to conservation. Tools and techniques for genomic analyses and manipulation have been developed for a wide range of model organisms, such as insects, rodents, plants, fish, and worms, as well as livestock like cows and pigs, and crops such as corn and soy. However, wild populations are generally much more genetically diverse than lab-adapted strains and domesticated organisms. Furthermore, it is precisely the genes that are responsible for adaptation that will vary the most. Mutation drives evolution, and organisms that have developed specific genetic variants more suitable to survival under particular selective pressures will eventually overtake those members of a population that cannot adapt. Without honing our understanding of how specific genes might affect survival in the face of environmental stresses, it may be difficult to know what variants to look for. This means that in order to assess whether a specific population is well equipped to deal with a particular environmental change taking place, the level of complexity will be significantly higher than if a retrospective analysis were conducted comparing different populations that are known to be adapted to the new conditions with those that are not. Identifying drought-tolerant individual plants within a wider population of plants required to prevent soil erosion or determining whether indigenous honeybees will rapidly move on if temperatures rise will have to be assessed before conditions change irreversibly. Tools and techniques for genomic analyses are critical to being prepared for irreversible environmental changes in ecosystems, and basic scientific investigations identifying candidate genes and understanding how specific mutations might alter structure and function are extremely important.

One issue that also adds significant complexity to conservation genomics is that often singling out specific species for analysis is not as useful as assessing complex mixed samples. In many cases, ranging from predator-prey relationships to other interspecies interactions, different species must coexist for ecosystem stability. One example of a complex mixed population relevant to other organisms are the soil bacteria that are at the base of many ecosystems. The microbiome (the microbial populations that exist together in one specific place and time) is actually quite well suited to genomic analyses. Identifying specific bacterial species through genotyping is quite straightforward, and these types of analyses can be applied to everything from an animal's gastrointestinal tract to the microbes associated with the root of a tree. One area where genomic analyses of mixed bacterial populations is proving extremely productive is in the analysis of feces to assess aspects of animal health and well-being, since stresses like disease and starvation will alter the fecal microbiome in predictable and reproducible ways. In a nutshell, if you want to know about the health of deer in a specific forest, just look at the bacteria in their poop.

Another way that genomics can be useful in conservation is in identifying related populations and species. The American bison was once on the brink of extinction but is now rebounding. Many bison are protected, but conservation efforts are complicated by the fact that these wild animals can interbreed with domesticated cattle that escape from grazing herds, producing a hybrid known as beefalo. Applying genomic technologies could prove to be extremely useful in assessing the relative proportion of domesticated cattle as opposed to wild bison both in DNA samples from individual animals and within different populations, and this could greatly assist in determining how to most efficiently apply conservation efforts.

Genomics is also proving useful in aquatic environments and providing insights for the conservation of marine species like sharks and whales. Although it makes perfect sense that by taking DNA samples from specific individuals, distribution patterns, migration, and genetic diversity can be tracked, this does require first finding the animals and obtaining DNA from them. However, similar to the way that circulating fetal cell-free DNA in the maternal bloodstream can be isolated for prenatal screening, researchers have been developing methods for sampling DNA directly from seawater. Environmental DNA (eDNA) can provide information assisting in applications ranging from detecting aquatic pathogens to studying biological diversity in specific areas, and assessing the genetic diversity of particular populations. Of course, identifying different species is different from tracking individuals, and the techniques being employed to apply the study of eDNA to conservation are still evolving. Currently, if you find a great deal of eDNA from one species in an area, it can be difficult to ascertain if it is from a large number of animals, or one releasing a lot of eDNA—for example, while giving birth. But all in all, application of the study of eDNA in conservation is a fascinating area with tremendous potential.

Also relevant to the topic of conservation genomics is a recent report that involved a specific series of criminal investigations into smuggling. In this case, though, it wasn't actually the criminal who was tracked through genomic technologies but rather the contraband that was being smuggled. Through physical inspection and DNA analyses of samples of ivory seized by law enforcement, scientists were able to determine that smugglers often shipped the two tusks taken from individual poached elephants separately in different cargo containers. This observation would suggest that by matching networks of tusk pairs across multiple seized shipments of ivory, individual groups of criminals could be linked to larger pools of evidence. Tracking the DNA allowed the researchers to determine that three major criminal networks were behind the thirty-eight different seizures of smuggled ivory that had been analyzed. Previous work in this area developed genotyping tools that were able to geographically map specific areas where poachers were obtaining the ivory through comparison of DNA obtained from ivory samples to a genetic reference map of African elephants. Illegal ivory trade is estimated to be responsible for the deaths of up to forty thousand elephants per year, and hopefully by combining these types of genetic analyses scientists will be able to assist law enforcement with investigations into the entire criminal network of poachers and smugglers.

EARTH BIOGENOME PROJECT (EBP)

One extremely ambitious initiative in the area of conservation genomics is the recently announced Earth BioGenome Project (EBP). Very broadly speaking, there are two main types of organisms on this planet, prokaryotes and eukaryotes. Prokaryotes are organisms like bacteria that do not have organelles inside them to compartmentalize chemical reactions and form discrete sites of molecular storage and organization. We are eukaryotes, which have intracellular organelles, and so are all other animals, plants, fungi, and single-celled protists. Eukaryotes represent nearly all life that isn't microscopic—and also much that is.

While eukaryotes share many genetic, structural, and functional characteristics, they also demonstrate tremendous diversity. There are potentially as many as ten million different eukaryotic species on earth, although only about 15 percent have been identified. A tiny fraction of these have been analyzed at a genetic level. Currently, the National Center for Biotechnology Information (NCBI) database only contains a few thousand eukaryotic genome sequences. What the EBP proposes to do initially is to sequence the genomes of representative members of the nine thousand or so different eukaryotic families. In evolutionary and genetic terms, families represent highly related groups that share most recognizable traits—for example, lions, tigers, and house cats are all members of the family Felidae.

These initial genome sequences will then serve as references on which to build and compare sequence data from all 1.5 million known eukaryotic species. The EPB is made up of a large international scientific team of ecologists, geneticists, and representatives from museums and biobanks that curate large collections of material that can be used for DNA extraction. In the initial phase, the focus will be on samples that have already been collected. Following this stage, new samples will be obtained in field expeditions.

The EBP will help us better understand biodiversity, which will in turn assist with conservation efforts. The EBP project may also help identify new beneficial natural products—such as medicines, many of which originally came from eukaryotic sources. Aspirin was first synthesized from willow tree bark, and penicillin is derived from a fungus. There are many other rationales behind this endeavor that have been suggested by the EBP, including improved agricultural and industrial processes, better pest control, fermentation, and rubber production.

The EBP has estimated that the sample collection and sequencing phases of the project will take ten years and cost $5 billion. Adjusting for inflation, this is roughly comparable to the cost of the Human Genome Project. Considering the revolutionary impact of the first human reference genome, the members of the EBP are confident their efforts will be transformative on a global scale.

ERADICATION

Although genomic technologies are able to aid conservation efforts and improve our understanding of how life on earth is changing because of human activity, alterations in climate, and abiotic considerations, there is another side of this coin that also requires attention—namely, applying these tools to eradicate unwanted organisms, populations, and species. Although the concept of eradicating a species might seem extreme, it may have considerable merit when it comes to mitigating the toll of human pathogens. In examples such as smallpox or polio, eradication campaigns have spared us untold amounts of human suffering. Although diseases such as these have generally been eradicated thanks to the painstaking work of medical professionals and large-scale epidemiological and immunization efforts that involved huge international collaborations, applying genomic technologies could, in some cases, make eradicating human pathogens more efficient and direct.

One major target for eradication assisted by genomic technologies could be what are referred to as neglected tropical diseases. These illnesses are mostly caused by pathogenic organisms that primarily exist in the developing world. Furthermore, many are not easily prevented, treated, or cured, at least not given the constraints of the environments in which they persist. Examples of these types of diseases include malaria, dengue fever, African sleeping sickness, and Chagas' disease. One key reason genomic technologies can help in eradicating some of these neglected tropical diseases is the fact that many originate not from bacteria and viruses but from eukaryotic parasites that follow complex sexual life cycles. Genomic means can then be applied to study the genetics of the pathogen. Bacteria and viruses can very easily change their genomes through mutation or swapping genetic material with others in a population, and sometimes even with their hosts. Eukaryotic parasites that employ sexual reproduction demonstrate patterns of genetic change and distribution in man ners much more similar to our own, which can make them much more easily traceable.

The tools of genomics can be applied to study these pathogens in the same way as they might be used in the context of conservation. However, rather than focusing on particular sensitive populations in order to save them, these tools would help to discover the open doors or low-hanging fruit that could pave the way for successful eradication strategies to be developed. Even with simple viruses and bacteria, genomic analyses can be much more powerful than standard binary diagnostic tests in both a clinical and epidemiological context. Rather than only telling you if a person is positive for a specific pathogen, by identifying which specific genetic variants the particular organisms causing that infection are carrying, genomic analyses can reveal details about aspects of that particular infection, such as whether an outbreak might be become more or less dangerous or if the organisms in question are sensitive or resistant to specific treatments.

In some diseases an alternative to targeting the pathogen is to focus on the so-called vector that introduces the disease-causing organism into the host. In many cases mosquitos carry disease, such as protozoan parasites and viruses, and humans can become infected once bitten. In a high-security lab in Italy, genetically modified mosquitos are currently being developed and studied that could one day lead to eradication of diseases like malaria, dengue, and Zika. These genetically modified mosquitos carry a genetic mutation that makes females unable to lay eggs and inhibits their ability to feed on human blood. As this type of mutation would not generally be advantageous in the wild, the scientists involved are applying a controversial technique to ensure spread of the mutation through the population. Termed gene drive technology, this method of genome editing increases the odds of inheriting a dominant mutation well above the 50 percent level dictated by Mendelian inheritance. Using CRISPR or other similar types of DNA-cutting enzymes, the transgene of interest can be spread from one chromosome to another and then into offspring and through the population. The researchers hope this genomic trick will be able to rapidly "crash" wild populations of mosquitos that carry deadly human disease.

But some have voiced significant reservations about deploying this technology in the wild and fear that release of these genetically modified organisms could lead to spread of the deleterious genetic variant outside the targeted population. Furthermore, unintended consequences could result from removing a population of insects from an ecosystem. Their absence could possibly lead to the rise of other species that could carry other diseases. Similarly, as issues such as predation and pollination involve complex networks of species, rapidly removing one component of a complex ecosystem might not be ultimately advisable.

Deploying the tools of genomics can help with conservation activities, particularly those aimed at understanding and potentially limiting the effects of climate change on fragile ecosystems. Many of these same techniques could be applied to understand and fight against the agents that cause human disease. Although deploying genomic technologies in the hope of altering the makeup of wild populations might seem ethically dubious, in many ways it is simply a more specific and targeted approach to assist efforts that have already been taking place to battle spread of disease.

PART V

Beyond the Genome

CHAPTER 23

The Basics of Epigenetics

The basic definition of epigenetics is any heritable change that does not rely on DNA sequence. Epigenetics generally involves chemical modifications to nucleotides, or DNA-binding proteins that alter the structure and function of chromosomes in ways that regulate gene expression. In many cases, these alterations actually change the structural shape of chromosomes and increase or decrease access to the transcription factors and polymerases that regulate gene expression.

What does DNA actually look like inside the nucleus of a cell? How is it organized? You may have seen images of chromosomes that look like little Xs, but that is only what chromosomes look like at one very specific phase of the cell cycle. During mitosis, when a cell is going to be duplicated it must copy all of its DNA. So the process involves going from one paternal copy of each chromosome and one maternal copy to two identical copies of each. These two "sister chromatids" are then linked together at around the halfway point along their length through a structure referred to as the centromere. The little X shapes are actually two identical copies of the same chromosome stuck together, and these align along the central axis of the cell that is about to be duplicated. The Xs are actually more like two parentheses, connected at their midpoints by the centromere:)·(. The two sister chromatids then separate from each other, during the final stages of mitosis, and one moves into one developing daughter cell while the other moves into the other daughter cell.

During normal cellular function, chromosomes don't typically look like little Xs. To make those shapes, the chromosomes must become significantly condensed—the DNA packed tighter than the 6 train at rush hour. Overall, the shape of chromosomes is very carefully regulated. Each chromosome has a different size, shape, and structural organization that can be regulated locally and globally—for example, by epigenetic means. If stretched out, the genomic DNA in each cell would be about two meters long, over six feet. Just before the cell splits in two, mitotic chromosomes are only a few micrometers (millionths of a meter) long. Even during the other more standard phases of the cell cycle, DNA must be condensed to an extreme degree. The nucleus of most cells is generally less than ten micrometers in diameter. Considering the sizes involved, one has to ask, How is it possible to get this huge mass of thread into such a tiny sewing kit? The answer is, through very careful organization.

There are proteins in the nucleus called histones, and DNA winds around them like little spools of thread. An entire chromosome doesn't wind around one histone, however. Rather, many different spools of DNA form around separate clusters of histones, at various places along the length of a chromosome, wrapped like long strands of hair around a series of curlers. This allows the DNA to be packaged very tightly and facilitates local unlooping—loosening or taking one curler out while leaving the others in place. The unlooping provides regulated access to the machinery that drives gene expression, such as transcription factors and RNA polymerase, because DNA that is tightly wound around histones cannot be easily expressed. During mitosis, when the chromosomes are packed into the tiny little X shapes, gene expression largely ceases. This is evidence that chromosome structure itself is closely associated with the ability to express genes from specific regions in the genome, and that by regulating the former you can control the latter.

The genes found in chromosomal DNA encode specific proteins with particular functions. In the nucleus DNA is transcribed to RNA by the RNA polymerase. Once transcription is complete and a fully processed messenger RNA (mRNA) is formed, it will then leave the nucleus. The mRNA is then translated to protein by ribosomes found in the cytoplasm. These are the basics of the so-called central dogma of molecular biology: DNA is transcribed to RNA, which is in turn translated into protein.

There are many ways gene expression can be controlled during the various steps in this process. However, the intermediate step of transcription of DNA to RNA is commonly used in regulating gene expression. By controlling the efficiency of transcription—for example, by regulating the ability of the RNA polymerase to access and bind target DNA sequences—gene expression can be precisely and specifically controlled.

The process of winding and unwinding DNA around histones represses and promotes gene expression, respectively, by regulating access to the RNA polymerase. This is a critical level of regulation for gene expression in the cell.

CONTROLLING GENE EXPRESSION

Interestingly, there are some mechanisms that control gene expression that can be carried over across generations, either from a parent cell to daughter cells through the phases of cell division, or even from parent to child. One example that will be described further below is that of gene silencing when epigenetic change in one generation can persist into the next. The mechanisms that are responsible for regulating how tightly DNA is wound around histones at particular locations are critical aspects of what are referred to as epigenetic regulation. These include adding small chemical groups to the DNA itself or the histones around which DNA coils in the nucleus, the hair and the curlers. Thus, these represent different ways apart from simple DNA sequence variation that heritable changes in gene expression can occur in cells and organisms.

As stated above, generally speaking, epigenetics is defined as heritable mechanisms that can be passed from one cell to another or one generation to the next that regulate gene expression apart from DNA sequence. If you want to take epigenetics to its most extreme, all influences that regulate gene expression and could be maintained across generations could be included. This would encompass transcription factors and proteins that bind enhancer, promoter, and repressor regions, among other modes of regulation that happen through hormones or any other molecules that can alter expression levels of different genes.

Even further afield is the broader concept of nongenetic inheritance (NGI). Epigenetics is one aspect of NGI, but theoretically NGI could include anything heritable that modifies an organism apart from DNA sequence alterations, even if it doesn't directly alter gene expression levels. This could include influences like learning, culture, and parenting. NGI is anything that could impart an advantage to the next generation under a specific set of environmental conditions—selective pressures, in other words. An example of NGI would be instructing children to eat a certain kind of berry that carries specific beneficial vitamins, and avoid berries that are toxic.

This perspective on NGIs can be taken further. It has been shown that male circumcision is protective against certain sexually transmitted diseases like HIV. In a traditional Jewish population, being uncircumcised would certainly have a negative effect on potential reproductive success, as not being circumcised would generally be a "deal breaker" in most observant Jewish relationships. One could make an argument that male circumcision among people of Jewish descent is an extreme example of NGI. Consider that a Jewish boy displays a certain physical trait, a circumcised penis, because of specific heritable conditions—the tenets of his parents' religion. A father who was circumcised has a son who is circumcised, and this continues on down the line. Of course, people can convert, and so being Jewish isn't exclusively linked to any particular genetic considerations. Similarly, individuals who are not Jewish can also circumcise their children, and do for religious, cultural, and medical reasons. Finally, individuals with Jewish ancestry can choose to participate, or not, in whatever cultural behaviors and traditions they wish. However, these exceptions do not necessarily invalidate the definition of NGI as *all causes of change in an organism that can be passed from generation to generation, either in cells or individuals, regardless of a direct effect on gene expression.*

NGIs aside, below we will focus on epigenetic mechanisms that do change gene expression, generally by directly modifying chromosomal structures, and can potentially be retained following cell division or sexual reproduction so that they pass on to future generations. The types of epigenetic mechanisms that chemically modify chromosomes and are more widely accepted and easily described include processes that directly regulate expression of particular genes, or groups of genes, in a controlled and predictable manner. In general, this definition of epigenetics includes two key types of chemical changes, either to the DNA itself or to the histone proteins around which the DNA winds.

DNA can be modified not only by alterations to the nucleotide sequence (e.g., a mutation that swaps an A for G) but also by chemical changes to the nucleotides themselves. This can alter the ability of DNA-binding proteins to interact at specific sites along a chromosome. The most commonly analyzed form of nucleotide modification involves adding a methyl group, a carbon and three associated hydrogens, to cytosine (C) nucleotides through an enzymatic reaction. Formation of 5-methylcytosine (5-MeC) in a DNA sequence is an example of an epigenetic modification associated with decreased gene expression in a targeted, specific, and localized way. Locations in chromosomal genomic DNA with 5-MeC modifications are associated with decreased gene expression potentially in part owing to alterations in binding of the site-specific transcription factors that regulate RNA polymerization. 5-MeC can be added or removed at particular chromosomal sites to regulate expression of specific genes; this might occur to reduce the local expression of proteins that slow down cell division in cells that have become cancerous or as part of the global changes in methylation during the stages of development and aging. Furthermore, the pattern of methylation in a particular cell is copied when cell division takes place, although the molecular details of how this happens in different contexts are still under investigation.

One example of how gene expression is regulated in this way is observed when comparing different tissues. Although all our cells share the same genome, different genes are expressed at different levels in different tissues and cells, and at different times. The genes in white blood cells and neurons are the same, but the transcription of those genes to RNA and then the translation of that RNA into protein—the expression of the genes—is what makes the different cell types unique and allows them to change over time. The expression levels of specific genes are quite different in the brain compared to the liver and the heart. Tissue-specific methylation patterns are believed to underlie these variable expression levels. Moreover, large amounts of 5-MeC found at particular regions can actually bring about gene silencing (where expression of specific genes is completely shut off).

GENE SILENCING

One extremely well-studied example of gene silencing occurs in what is referred to as parental imprinting. About 1 percent of human genes are expressed only via the maternal or paternal copy, rather than both. This proportion is consistent across the population and is tightly controlled and regulated by epigenetic means. The process of imprinting genes in this way seems to largely rely upon DNA methylation of specific genes and occurs in many, but not all, organisms that undergo sexual reproduction. Interestingly, maintaining epigenetic regulation through reproduction is complex as it is generally thought that DNA methylation is removed and reset between generations.

Imprinted genes tend to be involved in growth and development, and the process of imprinting genes follows a regular cycle. During the formation of a sperm or an egg, the methyl groups responsible for the previous imprints are removed and then reformed for all sperm or eggs to the paternal or maternal version, respectively. Thus, sperm and egg solely display the characteristic paternal and maternal imprints they had before fertilization. After fertilization and throughout growth, development, and adulthood, these methylations are maintained, and only the non-silenced copy of the gene is active and expressed.

If only the maternal copy of gene 1 is expressed, and the paternal copy is imprinted, and the paternal copy of gene 2 is expressed and the maternal copy is imprinted, then these different versions could be represented as gene 1(M) / gene 1*(P), and gene 2(P) / gene 2*(M) in all individuals, regardless of sex (* means imprinted and silent, or not expressed). If the individual is male, then all sperm should have gene 1* and gene 2, while if the individual is female all eggs should have gene 1 and gene 2*. This type of methylation is the mechanism that controls silencing to ensure that all paternally imprinted genes are silenced in all sperm and all maternally imprinted genes are silenced in all eggs. As sperm and egg carry only one copy of each gene, if a particular developing sperm contains gene 1(M), from the mother of the male, then that must become imprinted, as it will become paternal in the next generation and must be silenced. Similarly, if a sperm has gene 2*(M), also from the mother of the male, this must be deimprinted so that it will reflect the appropriate paternal pattern and be expressed after fertilization. These processes depend upon the enzymes responsible for methylation and demethylation. However, it also seems that other types of epigenetic mechanisms may play a role in imprinting, though this is less well documented and understood.

Another commonly assessed epigenetic mechanism that has been suggested to play a role in parental imprinting and several other biological processes involves modification of the histone proteins. Acetylation of histone proteins involves addition of an acetyl group, a carbon atom bound to an oxygen atom and a methyl group (one carbon and three hydrogen) for a total of two carbons, an oxygen, and three hydrogens. Histone acetylation causes DNA to unwind from around the spool and permits access to the machinery for RNA transcription such as transcription factors and RNA polymerase, promoting expression of nearby genes. Enzymes that demonstrate histone acetyltransferase (HAT) activity promote histone acetylation, which is associated with many types of situations where transcriptional upregulation is employed to stimulate gene expression. Many transcription factors that are activated following stimulation of cells with growth factors display HAT activity, and this opens up the chromosomal DNA, allowing binding to the promoter regions driving expression of genes involved in mitosis. Additionally, histone deacetylases (HDACs) are involved in chromosome compaction and down-regulation of gene expression through reduced levels of transcription. HDAC enzymes remove the acetyl groups from histones, and this allows the DNA to wind more tightly, reducing access for the transcriptional machinery.

HAT activity is like a mechanism that holds a book open, allowing readers to transcribe what is typed on a particular page into their notes. HDACs then close the book. In keeping with the book analogy, you could then consider DNA methylation, at least in the context of gene silencing, as redacting specific text in a book, crossing something out so it cannot be read, and removing methylation from a gene as highlighting or underlining specific words or sentences.

X CHROMOSOME INACTIVATION

Similar in some ways to parental imprinting, X chromosome inactivation in females is also a type of epigenetic change. Males have one X chromosome and one Y chromosome, and each sperm carries either one or the other. Females have two X chromosomes, and all eggs carry one. If a sperm carrying a Y chromosome fertilizes an egg, an XY male offspring will result. However, if a sperm carrying an X chromosome fertilizes an egg, each cell in the female offspring will have two copies of every gene on the X chromosome. To prevent overexpression of these genes, one of the X chromosomes is inactivated in the early embryo. This occurs via epigenetic mechanisms such as DNA methylation and histone deacetylation so that the DNA within the inactive X chromosome is tightly wound and completely inaccessible to the transcription machinery.

Chromosome X inactivation occurs very early in embryonic development. Much of the research in this area has been performed with mouse models, and although species-dependent differences may exist between the mechanisms that control X inactivation in mice compared to humans, it seems that both X chromosomes in all cells of an early female embryo become reactivated, and then randomly one of the two X chromosomes is inactivated within each cell. This change is then heritable through cell proliferation during growth and development. Once a particular X chromosome is inactivated in a particular embryonic cell, that same X chromosome will remain inactive in all cells subsequently derived from that cell throughout development into the adult organism.

The inactivated X chromosome was first observed about seventy years ago by Murray Barr, and it is still referred to as the Barr body. That being said, about a decade later it was English geneticist Mary Lyon who first proposed the mechanism of X inactivation. Interestingly, Lyon worked for a time in the laboratory of Conrad Waddington, who is generally credited as the founder of the field of epigenetics.

It should be noted that histones are not the only proteins that can be regulated through acetylation. The HATs and HDACs that can increase and decrease gene expression, respectively, through epigenetic modifications to histones, can also act on proteins besides histones. In some cases, these alternative targets for the machinery that regulates acetylation status include proteins with roles in cancer, so the roles of HATs and HDACs in health and disease might not be as simple as we once thought.

From fully extended linear chromosomal DNA to completely condensed mitotic chromosomes, many different levels of chromosomal structure exist. The material found inside the nucleus that includes chromosomal DNA and associated proteins such as histones is referred to as chromatin. Chromatin is found in different overall structural forms, including what is called heterochromatin, which is generally tightly packed and transcriptionally inactive, and euchromatin, which is more openly organized and includes sites of active transcription. In the case of X inactivation, the entire chromosome is essentially inactive heterochromatin.

However, beyond simple winding of DNA around histones, more complex degrees of chromosomal organization are now becoming better understood, such as long, open loops of chromosomal DNA and regions that appear to be attached to the nuclear periphery. Experimental techniques aimed at analyzing different types of chromosomal structures are being developed and applied to determine how gene expression can be regulated globally across a whole chromosome and more locally relative to specific regions and clusters of genes. However, from individual genes to whole chromosomes, many different mechanisms involved in structural regulation are being studied, and these operations are being analyzed for their roles in multiple biomedical processes ranging from embryonic development to complex human diseases.

Different regions of the genome can come into close proximity, causing regulatory consequences. Enhancer regions can be distant from the genes whose expression they control. The physical distance of two locations along a chromosome that has formed a structure such as a loop can be relatively small compared to the maximal potential separation gauged by DNA sequence. Chromosomes can fold into complex shapes that bring together positions that would be extremely far apart if the DNA strand was stretched out. Different chromosomes can be found in different positions within the nucleus; some are more centrally located, and others reside in the periphery. The organization of chromosomes can also change over time in response to certain factors like the stage of the cell cycle or the activation state of the cell. Different chromosomes can come into very close contact with their neighbors, and this proximity can be regulated by epigenetic mechanisms relevant to gene expression.

In addition to DNA and histone modifications, there are other epigenetic mechanisms that function solely at the level of RNA. There are classes of RNAs that can be exactly complementary to regions within mRNA. These so-called micro-RNA are, as the name implies, very short, approximately twenty or so nucleotides long. RNA, unlike DNA, is single stranded. So when a particular micro-RNA that matches a particular mRNA is expressed it can bind to the complementary mRNA, forming a double-stranded RNA. Enzymes in our cells are primed to attack double-stranded RNA and destroy it. Double-stranded RNA doesn't usually occur unless we are being infected by a virus with a double-stranded RNA genome, like the rotavirus responsible for terrible stomach problems. Similarly, destruction of double-stranded RNA formed through micro-RNA binding represents an epigenetic modification that has evolved to reduce gene expression by removing specific mRNAs.

Epigenetics is emerging as a key regulatory mechanism for diverse aspects of human health and disease, including learning, memory, cancer, and neurodegeneration. Depending on the targets and cellular context, the epigenetic mechanisms (e.g., DNA methylation and histone acetylation) can be rapid and reversible or become permanent. This can be very complicated to completely understand as the same general mechanisms can be employed in widely divergent contexts and result in different types of effects on gene expression. However, beyond tagging DNA with 5-MeC and histone acetylation there are several other less common forms of DNA and histone modifications that can modulate gene expression. The full complement of epigenetic mechanisms that might be involved in transcriptional regulation in health and disease is remarkably complex.

Epigenetics in Action

G enerally speaking, the role of epigenetic modulation for regulating gene expression in health and disease is not global—
at the level of the whole genome, like turning the volume on
your stereo up or down. Rather, transcription of specific genes or
groups of genes can be selectively up- or down-regulated by particular local epigenetic mechanisms, like the conductor telling the
violins or the timpani to get louder or softer. In cancer, for example, epigenetic mechanisms can result in the increased expression
of the oncogenes that promote cell division, or decreased expression of tumor suppressor genes that prevent inappropriate cell
proliferation. It seems that epigenetic changes can both be targeted through specific enzymes that regulate things like DNA
methylation or histone acetylation at particular locations, or they
can be more random, where the cells that proliferate more rapidly happen to be the ones with epigenetic changes that regulated
expression of genes involved in cancer in this way.

Several inhibitors of DNA methylation and histone deacetylation have been approved for clinical use as drugs, particularly for treatment of different forms of cancer. Azacitidine and
decitabine are drugs that inhibit DNA methylation that are both
currently used as treatments for certain types of leukemia.

As both DNA methylation and histone deacetylation are
involved in decreasing gene expression, treatment with these
drugs should in theory increase expression of genes that inhibit
cancer progression. But the reality is seldom so simple. There

arc also some histone acetyl transferases (HATs) that seem to be involved in promoting expression of genes associated with multiple forms of cancer and neurodegenerative disease. Some HAT inhibitors, which would result in decreased expression of certain genes, are also emerging as promising potential therapeutic leads in the fight against a range of diseases with an epigenetic component. Furthermore, site-specific alteration of epigenetic modifications through innovative molecular means, such as CRISPR-dependent approaches, are being developed to selectively silence particular genes.

Cancer is not the only disease associated with epigenetic changes gone awry. In the brain epigenetic mechanisms seem to be involved in learning and memory. It is therefore not surprising that epigenetic modifications regulating expression of particular genes in the brain have been strongly linked to neurological diseases and neurodegeneration. An important point to consider is that neurons in the adult organism generally do not undergo cell proliferation. Although modifications associated with epigenetic regulation of gene expression certainly occur in the brain, it does not seem as though they pass directly from one cell to another. As will be described below, the question of whether epigenetic changes to the human brain—for example, after extreme trauma—can pass from one generation to the next through sexual reproduction is an intense area of current research linking psychology, biology, and even world history.

ALTERATIONS IN THE BRAIN

Specific experiences can lead to epigenetic modifications in the brain that seem to underlie the formation of memories. Similarly, loss of memory like that seen in the case of Alzheimer's disease (AD) is strongly associated with epigenetic changes. The direct causes of these changes are quite hard to determine as AD is a complex disease that involves multiple genetic and environmental factors. Huntington's disease (HD), which is caused by a single dominant mutation, is a more straightforward disease, genetically speaking, and also seems to involve epigenetic changes in the brain. Dramatic alterations in epigenetic status can be observed after a stroke, which causes reduced blood flow to parts of the brain and significant neuronal cell death. Specifically discerning the extent to which epigenetic changes seen in different pathological situations in the brain might be the cause or the result of cell death and damage is not a cut-and-dried task. Although AD and HD are progressive diseases, in the case of stroke different phases are observed, from the original period of acute damage to gradual, and hopefully complete, recovery. The balances between genetic and environmental, or behavioral, causes for AD, HD, and stroke, are all very different. Although AD likely has genetic components leading to predisposition, as a dominant genetic disease HD is unavoidable if the mutant variant is inherited. Numerous genes likely contribute to relative risk for stroke, as evidenced by the large number of SNPs identified as more common in stroke patients, along with clear significant effects depending on lifestyle. However, given that stroke is a traumatic event, rather than a longer-term progressive process, simply observing increased or decreased methylation at a single fixed time after a stroke might not be extremely informative.

Drugs that alter epigenetic regulation of gene expression show clear effects in the brain. In rodent studies HDAC inhibitors seem to promote memory formation, suggesting a role for increasing gene expression by histone acetylation in creating new memories. Similarly, humans with genetic mutations that reduce HAT activity are more likely to display learning disabilities. Thus, histone acetylation, and therefore increased gene expression, seems involved in promoting learning and memory. Furthermore, as altered DNA methylation patterns are a common occurrence in diseases like AD and HD, drugs that target the epigenetic machinery that regulates DNA methylation are currently being developed as potential therapeutic agents for a variety of neurological diseases. HDAC inhibitors are already being employed to treat epilepsy. Since the brain is isolated from the general circulation by the so-called blood-brain barrier, any drugs developed and applied to target the epigenetic machinery in neurological disease must first be able to efficiently access the brain, commonly through chemical modifications that permit them to cross the blood-brain barrier.

THE ENVIRONMENT AND EPIGENETICS

Another example where epigenetic mechanisms seem to be involved in regulating gene expression is in response to toxic chemicals in the environment such as pesticides employed in industrial agriculture, gardening, and home pest control. In some cases, pesticides seem to cause effects apart from inducing mutations through alteration of genomic DNA that can potentially cause unintended issues in humans following inadvertent exposure.

Most pesticides approved for current usage in the United States are not believed to cause mutations in genomic DNA, yet evidence is emerging that they may alternatively function in humans via epigenetic mechanisms. Hematological malignancies (blood cell cancers) represent around 10 percent of new cancer diagnoses, and these seem to involve complex alterations in epigenetic gene regulation and have been associated with pesticide exposure.

The pesticide DDT, which is banned in the United States, is still employed in some countries in the world. DDT has been associated with altered DNA methylation in the brains of rats, among other toxic effects. Some pesticides can cause lifelong sexual and reproductive effects in rodents following exposure during early development. Arsenic, which historically was a component of many different pesticides, is also associated with changes in methylation. Paraquat, an herbicide, has been associated with inhibiting HDAC activity, hence increasing acetylation of histones.

Epigenetic changes associated with environmental factors are not limited to toxic chemicals like pesticides. There seem to be a multitude of external stimuli that can trigger epigenetic modification of gene expression. Some evidence suggests these changes might be able to be inherited from one generation to another. Recently, there has been a great deal of publicity around events described as "intergenerational trauma." These include children who were in utero during wartime or famines. However, the bar should be quite high in order to define these claims as being truly directly epigenetic in nature.

There are many examples of traumatic events leading to a clear difference in health and well-being in the next generation, ranging from metabolic issues such as obesity to neurological and psychological problems like schizophrenia. But the extent to which these fascinating, and often tragic, occurrences might be truly epigenetic in nature is not currently clear. Children born to mothers who were pregnant during times of famine and war have been shown in repeated studies to fare worse than others of the same generation and are often observed to suffer from a range of physical and psychological ailments.

Epigenetics has been suggested as a potential explanation for the finding of the remarkable health issues found in children born immediately after the Dutch Hunger Winter of World War II, or from mothers who were pregnant during the famines that took place in China around 1960 or in the Ukraine in the early 1930s. Some claims of intergenerational trauma, though, don't necessarily hold up to the generally accepted definition of epigenetics.

If the child was already in the womb when the trauma was sustained, then it doesn't hold that the trauma was truly inter-generational. If a pregnant woman were exposed to some sort of toxin that was passed on to the fetus and caused cell death or organ dysfunction, this would not necessarily be epigenetic in nature. Strictly speaking, things like fetal alcohol syndrome might not be due to altered epigenetic regulation of gene expression, from a mechanism apart from DNA sequence changes that passes directly from parent to child, or through the process of cell division from one cell to another. It is possible that any alterations observed in DNA methylation or histone acetylation could just as easily be the result of the toxic effects, rather than the cause. They could have originated directly within the fetus, rather than from maternal transfer.

In several cases of supposed intergenerational trauma the alleged epigenetic effects involve neurological and psychological issues, ranging from fear responses to anxiety and schizophrenia.

However, neurons in the brain generally do not undergo cell division following development. The trauma would need to have occurred while significant cell division was occurring in the fetus for these traumas to be technically defined as epigenetic, or more simply via parental gametes into the next generation.

A few studies have recently emerged that may be providing evidence for intergenerational inheritance of trauma. Mice that are exposed to a specific scent at the same time as an electric shock learn to fear that particular odor, in a classical Pavlovian response. Mouse pups born to fathers who were conditioned in this way also seem to demonstrate a fear response when exposed to the scent. This fear response even persists if the second-generation mice are conceived through IVF from isolated sperm from the father, and the effect continues to be seen one generation later, in the grandpups of the original mouse.

The authors of this study then went further into the molecular mechanism that was potentially behind their observations. They found that the gene encoding the receptor for the odor they employed in the fear conditioning was methylated less in the pups derived from the mice that experienced the odor-associated trauma than in mice whose fathers had not been conditioned in this way. The model that flows from this work is that reduced methylation of the receptor gene would lead to high expression of the receptor and increased sensitivity to the odor. These results would seem to potentially satisfy the criteria for true direct transgenerational epigenetic effects.

More recently, a study in humans similarly raised the question of whether trauma can pass from one generation to another in this way. Through analysis of health records, it was determined that sons born to fathers who were prisoners during the Civil War, an extremely traumatic experience, had significantly worse health outcomes, as measured by relative risk of death starting after age forty-five, than sons born to fathers who did not experience this extreme form of trauma. This study demonstrated that the observed differences were not the result of other factors, such as family socioeconomic conditions.

What the Civil War POW study and the mouse odor experiment have in common is that the trauma seems to have been passed through sperm along the paternal line, and the affected offspring were not exposed in the womb. If true, these observations would support an epigenetic explanation. However, the POW study does not necessarily describe a specific epigenetic mechanism, the exact genes that are alternatively expressed because of particular chromosomal modifications. These observations are more hypothesis generating than truly providing definitive conclusions on the mechanisms of transmission. However, they do lend strong support for entertaining the possibility of intergenerational trauma in humans occurring via epigenetic means.

EPIGENETICS, HEALTH, AND DISEASE

Opening up beyond the standard definition of epigenetics, many other interesting observations are emerging in this area. Transcriptional activity is not the only thing that can be regulated by chromosomal modifications associated with epigenetic pathways. Telomeres are regions at the very ends of chromosomes that do not contain genes, sort of like those little holders you stick into the ends of a piece of corn on the cob. Telomeres generally shorten over time following rounds of cell division, and shortened telomeres are associated with aging and the loss of the capacity for cell division, termed senescence. Telomeres can be modified by epigenetic mechanisms such as DNA methylation and histone acetylation in both health and disease. How this is all regulated isn't fully known yet, but new evidence is suggesting that the epigenetic state of telomeres seems to be critical to the regulation of telomere length.

Telomere length is maintained in cancer cells. They do not shorten progressively in the same way as noncancerous cells, and the cancer cells continue to divide. Recently, it has become evident that some anticancer drugs that target epigenetic pathways can lead to telomere shortening and senescence. The exact roles for different epigenetic modifications in telomere length in normal and cancer cells are not currently very well understood, but this is one example of how epigenetic modifications might play a role in disease processes, as well as in aging.

Another example of epigenetic changes relevant to aging involves progressive methylation of chromosomal DNA over time. Generally speaking, methylation increases over time and is associated with aging. In fact, assessing DNA methylation can help estimate the age of an individual quite accurately, and it can be measured very simply from a blood sample. Furthermore, people who live to extremely old ages seem to display slowed progressive DNA methylation changes over time when compared to people with shorter life spans. This global methylation increase over time may be directly involved in aging. This raises the question of how we characterize biological versus chronological aging. So it might be true that what matters most is how old we feel!

In addition to this level of uncertainty in extrapolating DNA methylation status as a marker for chronological age, there are potential issues such as the possibility of testing samples from individuals with diseases like cancer that are associated with altered methylation. Commercial tests for assessing the "methylation clock" are currently being marketed. However, the error rates in such tests limit their potential usefulness. This is an active area of research experiencing rapid improvement: the average accuracy of these tests in pinpointing chronological age has already improved from about three years to within approximately one year of true age, although there might always be a few "outliers" whose methylation clocks function differently—and thus are not appropriate candidates for testing in this way.

Overall, epigenetics is emerging as a critical aspect of development and represents a promising avenue for enriching our understanding of health and disease. Most of the diseases that have plagued mankind—from cancer to heart disease and type 2 diabetes—seem to have some inherited aspects yet also depend upon environmental influences such as diet. Because genomic analyses such as GWAS often do not provide anywhere near a complete understanding of the heritable components underlying disease, it would be remiss not to tap epigenetics to fill these gaps in our understanding.

One analogy that has been proposed to explain the role of epigenetics in human disease is to so-called dark matter. Astrophysicists have tried to compare the amount of matter in the universe through both optical means (light) and gravitational measurements (mass). They have found that something like 85 percent of the matter in the universe seems to be completely invisible. Thus, they have postulated dark matter as an explanation. Epigenetics may likewise be the invisible explanation for why genetic variations such as SNPs do not explain the complete heritable aspects of the vast majority of human disease.

New technologies to measure epigenetic modifications are being developed. Techniques for measuring epigenetic changes in real time, rather than from fixed end points, and in single cells are a very hot area of current investigation. Similarly, analytical platforms are being developed that can directly measure epigenetic modification of specific chromosomal regions and the expression of genes found at those locations. But since we know that the nucleus is not the only source of genetic material, it should come as no surprise that interest in the role of mitochondrial DNA in reproduction, heath, and disease is also growing dramatically.

CHAPTER 25

Mitochondrial DNA

Mitochondria are often referred to as the powerhouse of the cell. They are where the universal currency of energy, adenosine triphosphate (ATP), is generated. Ironically, although often referred to in dynamic and exciting terms in the popular press, to many students of biochemistry, the mitochondrion is just a big bag of boring enzymes. This is because the many intersecting metabolic pathways required to convert glucose to ATP are quite convoluted and complex to fully understand, so it is tempting to view them as a branching network of components with little underlying rhyme or reason.

Glycolysis is the first phase of the process and takes place in the cytoplasm of cells. This involves the chemical breakdown of the sugar glucose to form a few molecules of ATP and the starting materials for the Krebs cycle. The next phase takes place inside mitochondria and creates a little bit more ATP. However, the Krebs cycle also sets up the conditions required for the so-called electron transport chain. This is the main event, which is how the bulk of ATP is generated. In total, thirty-six molecules of ATP are generated per glucose molecule, and this can go on to power a range of chemical reactions throughout the cell.

Mitochondria look like bacteria and share similar structural organization. The endosymbiont hypothesis originally proposed by Lynn Margulis suggests that early single-celled organisms either engulfed or were intracellularly colonized by a bacterium, and this led to a mutually beneficial coexistence, or symbiosis, where the host cell was able to make use of products of bacterial metabolism, such as ATP, and the bacterium was sheltered from the dangers and stresses of the world outside the host cell.

One piece of evidence supporting the endosymbiont hypothesis is that mitochondria maintain a small circular chromosome containing a total of thirty-seven genes. Thirteen of these genes encode proteins, including several components of the electron transport chain. The rest are involved in the production of RNA that is important in gene expression, but do not actually become proteins themselves. Only a total of about sixteen thousand base pairs are present in the mitochondrial genome, compared to the six billion in the complete diploid human genome.

Like the chromosomal DNA found in the nucleus, mitochondrial DNA must be tightly packaged to fit inside a very small space. Although there are no histones associated with mitochondrial DNA, there are DNA-binding proteins that appear to be involved in regulating compaction and access to the transcription machinery. In nuclear DNA each cell contains two copies of each chromosome, but there are generally thousands of copies of mitochondrial DNA per cell, as every cell has many mitochondria each containing several copies of the circular chromosome. This permits robust expression of the genes encoded by mitochondrial DNA in each cell.

Mutations in mitochondrial DNA are common, and although the presence of many distinct copies of the mitochondrial chromosome can reduce the overall effect of some rare mutations, it can also lead to debilitating and potentially fatal mitochondrial diseases. Leigh syndrome, for example, is a complex neurological disease that can result from mutations in mitochondrial DNA. The overall process of mitochondrial DNA replication and inheritance is quite complicated and not completely understood. What is known is that all mitochondria, and mitochondrial DNA, comes from the mother. Human sperm contribute only nuclear DNA during fertilization. This both simplifies matters in certain ways, as all mitochondrial DNA comes from one source, with little chance for variation, and creates some significant potential complications owing to the relative lack of genetic diversity.

Like all diploid cells in a human, the precursor cells responsible for forming oocytes (immature egg cells) in females contain a large heterogeneous mix of mitochondria with different numbers of copies of the mitochondrial genome, as well as potentially different specific polymorphisms and mutations. Different oocytes can have unique complements of mitochondrial DNA depending on the specific mitochondria received. In this way, a woman who is a carrier for a genetic disease caused by a mutation in mitochondrial DNA can have some children who are unaffected, and others who have the disease. As this does not follow conventional patterns of Mendelian inheritance, which requires one maternal and one paternal copy of each gene, predicting the outcomes of these situations can be extremely complex. So both genetically and medically, understanding, preventing, and treating diseases that originate from mutations in mitochondrial DNA is a major challenge.

Mutations in mitochondrial DNA can cause debilitating human disease, including a number of rare neurological disorders such as Kearns-Sayre syndrome, Leber's hereditary optic neuropathy, and, as mentioned above, Leigh syndrome. Thus, innovative techniques have been developed to assist would-be mothers with a significant risk of having children affected by mitochondrial dysfunction. The medical interventions include injecting the ooplasm (the cytoplasm from an oocyte) donated by a woman with no risk of carrying mutations in mitochondrial DNA, into an oocyte collected from the affected would-be mother before *in vitro* fertilization (IVF), which is the combination of sperm and egg (oocyte) in the laboratory before implantation into the mother. The ooplasm injection allows for healthy mitochondria without mutations in mitochondrial DNA to be introduced into the fertilization process. Generally, the IVF technique of intracytoplasmic sperm injection is employed simultaneously with the ooplasm injection, and in this way the healthy mitochondria and the sperm are introduced together into the egg.

Another promising solution involves removing the nucleus from one fertilized egg formed from a donor, and adding the nucleus from the affected would-be mother. However, this technique involves first creating two successful zygotes (fertilized eggs, or one-cell embryos) via IVF, which adds considerable complexity to the process. This nuclear transfer technique is similar to the cloning methods that were employed to generate Dolly the sheep, Snuppy the dog, and others. The similarities to the procedures that could be used for human cloning can make some people very uncomfortable.

The method for assisted reproduction for would-be mothers who are carriers of mutations in mitochondrial DNA that appears to be emerging as the optimal choice is sort of a hybrid of the two preceding techniques. Basically, the nuclear chromosomal DNA is removed from a donor egg and replaced with the nuclear genome isolated from an egg from the would-be mother, and then this hybrid egg undergoes IVF. In this method all the mitochondria and mitochondrial DNA in the resultant child will come from the donor, but the complete nuclear genome comes from the parents. In some ways, by using this technique, genetically speaking, the baby actually has three parents.

These techniques, which have been possible for many years, are referred to as mitochondrial replacement therapy (MRT), although this name does not necessarily reflect all the cellular components that can be contributed by the donor. In the United States the FDA strictly prohibits MRT. This national moratorium pertains to both experimental application in any MRT research that might result in a live birth of a human child with three parents and to any clinical application. It is currently unclear exactly what the legal, ethical, and biological implications might be of having a child with three parents, two who contributed nuclear DNA and one that provided mitochondrial DNA. However, this has not stopped doctors in other countries from helping would-be parents through MRT techniques.

Beyond cases where the goal is to avoid mitochondrial disease, these procedures have also been shown to be potentially helpful as a general infertility treatment. This is particularly relevant to cases where the inability for a couple to conceive seems to stem from issues with the oocytes. For example, advanced maternal age can be associated with the presence of nonviable eggs and this often seems to be linked to issues with maternal mitochondrial function. From experimental studies to clinical application to avoid mitochondrial disease, as well as assisted reproduction in general, this is a complex area with many different facets.

When IVF fails because of nonviable eggs, donor eggs are the only alternative currently available. However, this means that the nuclear and mitochondrial DNA in the offspring will not be that of the mother. MRT produces children where the genome is a combination of the mother and father, apart from the mitochondrial DNA, which comes from the donor. MRT to avoid transmitting mitochondrial disease is legal in the UK, and there are certain doctors worldwide who are employing MRT to assist couples to conceive who have failed multiple rounds of IVF because of issues with oocyte viability.

Beyond adoption or using donor eggs for IVF, there are few options for women with high risk of having children with mitochondrial disease or who cannot generate viable eggs. For a woman with mutations in some copies of the mitochondrial genome, it can be extremely difficult to know the potential outcome of any specific pregnancy. While it is possible to genetically screen early embryos for the presence of mutations in mitochondrial DNA, given the complex nature of mitochondrial inheritance, there is no fixed threshold that can predict absolutely whether a child at risk will in fact suffer from a mitochondrial disease. Furthermore, female children born after this screening would still carry the risk of transmitting mitochondrial disease to their children. Interestingly, given the metabolic requirements of early embryonic development, it may well be that some effects of mutations in mitochondrial DNA might actually prevent pregnancy from progressing. Although this doesn't necessarily help with the problem of improving reproductive success in women who carry these mutations, it might ultimately reduce the risk of giving birth to children affected with mitochondrial disease.

Although many women who carry potentially disease-causing mutations in mitochondrial DNA, and those who otherwise cannot generate viable eggs, want to have children—and a potentially effective option seems to exist—legal, ethical, and regulatory boundaries are preventing many couples from becoming parents. The concerns that have led to these regulations are certainly valid. However, as we have seen with the potential application of CRISPR to human genome editing, creating rational and reasonable international guidelines and regulations for applying genomic technologies to modifying the human species is not simple in any way. The question of what the future will bring is a big one. As individuals and as a species, we seem to be hovering on the brink of unlocking myriad novel ways in which genomic technologies may revolutionize many facets of life as we know it.

CHAPTER 26

The Future

What will the future hold? Will we soon be living in a world where screening of circulating fetal DNA from developing embryos by whole genome sequencing tells us everything we want to know about unborn children—from their potential risk for developing a range of diseases to their athletic prowess, intellectual capability, and earning potential? Will an international global genomic database provide each individual with a personalized family tree going back generations? Will techniques for genetic engineering such as genome editing allow us to design and optimize the food we eat, our pets, or our children? What are the critical technological, ethical, legal, regulatory, and economic limitations preventing the widespread deployment of innovative genomic technologies?

One key question moving forward into the future of genomic and personalized medicine concerns which technologies might ultimately become part of the standard of care in clinical applications. Currently, many assessments of individual risk for diseases with genetic components employ quantitative SNP arrays. But as the cost of whole genome sequencing drops, this technique might become more of a clinical reality, rather than remaining primarily a tool for research studies. Similarly, exome sequencing could become a standard method for screening individuals for mutations in coding sequences, or be demoted to a half measure if whole genome sequencing completely overtakes it.

Areas such as epigenetics and mitochondrial DNA clearly demonstrate that genomic sequencing cannot be the sole focus for understanding heritable disease. Similarly, techniques like exome sequencing require significant sample preparation that might not be available for all types of applications, especially beyond the human species. Furthermore, each of these techniques presents different challenges when it comes to data processing, transfer, storage, and analysis.

Historically, genetic diseases have been classified either as monogenic, which means that they are caused by a single mutation in a particular gene, or polygenic, where polymorphisms in multiple genes combine to cause disease. Monogenic diseases are often recessive, which means the individual must inherit two mutated copies, and include diseases such as cystic fibrosis. These variants are rare and display a very strong effect, leading to a clinically defined disease, and can be identified by DNA sequencing. However, not each and every individual that inherits these mutations will display the exact same disease symptoms. It seems as though other considerations, such as variants in other genes and the environment, can modify the disease severity.

On the other hand, polygenic disease often involves polymorphisms with subtler effects, which are more common in the population. Techniques like GWAS use SNP arrays to characterize the different polymorphisms that are inherited along with the disease. Furthermore, influences like environment and behavior are often involved in determining whether someone who is at risk of developing a polygenic disease will become sick or not. These types of diseases include disorders like type 2 diabetes. However, often even large-scale GWAS studies of huge populations, with as many as hundreds of thousands of individuals, cannot explain all the heritability associated with a disease. It is, of course, possible that epigenetics plays a significant role in explaining some of this "missing heritability," but one interesting perspective that seems to be emerging in the field of human medical genetics is that most human diseases with a genetic component actually emerge from a mixture of rare mutations with relatively strong effects and more common polymorphisms with weaker impact.

A potential procedure that seems to be gaining traction is to perform GWAS to characterize the common variants associated with a disease, and then follow up with more detailed studies, such as whole genome sequencing on a subset of the overall sample to identify rare and perhaps novel mutations. However, how the population for the detailed analysis is chosen is critical. In many cases, although studying genetically homogeneous populations can result in data with the highest statistical significance, this also means that the observations made in a study of this type cannot easily be extrapolated to other more genetically distinct groups.

If cost and the big data problem were not concerns, whole genome sequencing would most likely be the preferred method of choice for resolving most questions about genetic issues. Looking into the future, this might well come to pass for nearly all applications. However, error rates in whole genome sequencing remain a significant concern, and newly identified mutations should still be confirmed. With whole genome sequencing data, it is quite easy to focus on specific genes, groups of genes, or only the exome, or perform a SNP analysis through purely computational means. However, for now, the sheer size and complexity of whole genome sequencing data, plus the fact that it is still a challenge to sequence repetitive regions of the genome, make this technique not yet ideal for all questions and applications.

Another variable beyond which genomic technique to apply involves what types of samples to obtain and analyze. Preimplantation genetic screening in IVF and testing for circulating fetal DNA are already revolutionizing reproductive medicine. Analysis of circulating DNA is poised to play a significant role in understanding the progression of cancer, and responses to different therapies. Other applications for circulating DNA are certainly on the horizon, but analysis of circulating RNA may ultimately prove to be more informative, as this will tell us about gene expression levels, not just sequence.

DNA testing can represent more of a binary identification: either the embryo contains a third copy of chromosome 21 and will thus develop Down syndrome, or not. The extent of transcription of specific genes can be functionally linked to specific diseases. There are particular genes that are associated with metastatic cancer, and are thus expressed when tumor cells spread throughout the body. It might be possible to monitor circulating RNA derived from cancer cells to create a molecular signature of the genes that are being expressed to determine whether the disease is metastatic.

There are limitless potential applications for circulating DNA or RNA testing, as well as many other innovative genomic technologies currently being developed. It is possible that widespread whole genome sequencing will be facilitated by instruments that are fast, accurate, cheap, reliable, and equally applicable to all types of sequencing applications. This new generation of DNA sequencers may surpass the combined limitations of the currently available solutions and emerge as a single gold standard. Similarly, advancements in storage solutions, transfer bandwidth, and AI might once and for all solve the big data problem in genomics.

In terms of genetic engineering and genome editing, it seems as though we are poised to enter a new age where certain human diseases can be eradicated permanently, and improvements to all aspects of food production can be easily brought about. In our well-founded enthusiasm, however, we must not allow ourselves to blithely overlook darker alternatives. Perhaps genomic manipulation will run amok, and those with power, means, and access to genomic technologies will breed humans to exhibit specific traits viewed as desirable according to arbitrary, prejudiced standards for defining success in life. Ultimately, the tools of genetic engineering might turn out to be a Pandora's box that inflicts severe heritable damage on our species, or others. The worst fears of the anti-GMO activist could come to pass, or we could all find ourselves suspects in law enforcement's global criminal database.

I choose to have faith in the scientific method, and in the legal, regulatory, and ethical frameworks that have been developed and will be created to protect us from these dismal imaginable outcomes. It was over sixty-five years ago that the structure of DNA was first described, and almost fifty years since the first

DNA sequence was determined. The Human Genome Project was completed nearly twenty years ago, and the advancements in genomics have been accelerating at a greater pace. Genomics is already making drugs safer, decreasing the use of dangerous invasive tests, and opening up new worlds of possibility for dealing with climate change, overpopulation, and persistent human disease. The future of genomics looks to be one of infinite possibility where efficient application of technological innovation can benefit us all.

ACKNOWLEDGMENTS

I would like to thank Doug Cook, Elaine Mardis, Beth McNally, and Steve Quake for offering their time and expertise. I would like to thank Shahrnaz Kemal for critically reading the manuscript. I would like to thank my wife, Ema, for her comments on the manuscript and the many hours of interesting discussion on these topics. Finally, I would like to thank my agent, John Willig, the whole team at BenBella Books, and especially my editor, Sheila Curry Oakes, without whom this project would not have been possible.

REFERENCES

Below is a list of resources that were consulted during the process of researching these topics and may be helpful for gaining more insight. These publications include a few books and websites, as well as a number of articles in the general press, scientific editorials, review articles, and primary research papers. These are organized by section as some may be relevant to multiple chapters.

Introduction and Part I

1. Amanda L. Ogilvy-Stuart and Stephen M. Shalet, "Effect of Radiation on the Human Reproductive System," *Environmental Health Perspectives* 101, no. S2 (1993): 109–16.
2. James D. Watson and Francis H. C. Crick, "Genetical Implications of the Structure of Deoxyribonucleic Acid," *Nature* 171, no. 4361 (March 1953): 964–67.
3. Ferris Jabr, "How Beauty Is Making Scientists Rethink Evolution," *New York Times*, January 9, 2019.
4. Francis S. Collins, Victor A. McKusick, and Karin Jegalian, "Implications of the Genome Project for Medical Science," National Human Genome Research Institute, last updated March 29, 2012, www.genome.gov/25019925.
5. Rowan D. H. Barrett et al., "Linking a Mutation to Survival in Wild Mice," *Science* 363, no. 6426 (2019): 499–504.
6. James D. Watson and Francis H. C. Crick, "Molecular Structure of Nucleic Acids: A Structure for Doxyribose Nucleic Acid," *Nature* 171, no. 4356 (April 1953): 737–38.
7. Cassandra Willyard, "New Human Gene Tally Reignites Debate," *Nature* 558(2018):354-355.
8. Charles Darwin, *On the Origin of Species* (London: John Murray, 1859).
9. José María Izquierdo et al., "Regulation of Fas Alternative Splicing by Antagonistic Effects of TIA-1 and PTB on Exon Definition," *Molecular Cell* 19, no. 4 (2005): 475–84.
10. Carl Zimmer, "Scientists Created Bacteria with a Synthetic Genome. Is This Artificial Life?," *New York Times*, May 15, 2019.

11. Kwang-Pil Ko et al., "The Association Between Smoking and Cancer Incidence in BRCA1 and BRCA2 Mutation Carriers," *International Journal of Cancer* 142, no. 11 (2018): 2263–72.

12. Joshua Z. Rappoport, *The Cell: Discovering the Microscopic World That Determines Our Health, Our Consciousness, and Our Future* (Dallas: BenBella Books, 2017).

13. Richard Dawkins, *The Selfish Gene*, 4th ed. (New York: Oxford University Press, 2016).

14. Ed Yong, "The Wild Experiment That Showed Evolution in Real Time," *Atlantic*, January 31, 2019.

15. Julius Fredens et al., "Total Synthesis of *Escherichia coli* with a Recoded Genome," *Nature* 569, no. 7757 (2019): 514–18.

Part II

1. Antonio Regalado, "All the Reasons 2018 Was a Breakout Year for DNA Data," *MIT Technology Review*, December 29, 2018.

2. Paul Cliften, "Base Calling, Read Mapping, and Coverage Analysis," in *Clinical Genomics*, ed. Shashikant Kulkarni and John Pfeifer (New York: Academic Press, 2015), 91–107.

3. Adi Gaskell, "Big Data and Genomics," *Huffington Post*, September 13, 2017.

4. Jon Markman, "Big Data Is Remaking Big Pharma," *Forbes*, August 10, 2018.

5. Zachary D. Stephens et al., "Big Data: Astronomical or Genomical?," *PLoS Biology* 13, no. 7 (July 2015): e1002195.

6. Alice Chien, David B. Edgar, and John M. Trela, "Deoxyribonucleic Acid Polymerase from the Extreme Thermophile *Thermus aquaticus*," *Journal of Bacteriology* 127, no. 3 (1976): 1550–57.

7. Frederick Sanger, "Determination of Nucleotide Sequences in DNA," Nobel lecture, December 8, 1980.

8. Elaine R. Mardis, "DNA Sequencing Technologies: 2006–2016," *Nature Protocols* 12, no. 2 (2017): 213–18.

9. Frederick Sanger, S. Nicklen, and A. R. Coulson, "DNA Sequencing with Chain-Terminating Inhibitors," *PNAS* 74, no 12 (1977): 5463–67.

10. Jessica Baron, "Don't Believe the Headline Hype When It Comes to Genomics Research," *Forbes*, October 22, 2018.

11. Steven H. Wu et al., "Estimating Error Models for Whole Genome Sequencing Using Mixtures of Dirichlet-Multinomial Distributions," *Bioinformatics* 33, no. 15 (2017): 2322–29.

12. Jeffrey D. Wall et al., "Estimating Genotype Error Rates from High-Coverage Next-Generation Sequence Data," *Genome Research* 24, no. 11 (2014): 1734–39.

13. Amanda Warr et al., "Exome Sequencing: Current and Future Perspectives," *G3: Genes, Genomes, Genetics* 5, no. 8 (2015): 1543–50.

14. Louis Papageorgiou et al., "Genomic Big Data Hitting the Storage Bottleneck," *EMBnet.journal* 24 (2018): e910.

15. Barry Moore et al., "Global Analysis of Disease-Related DNA Sequence Variation in 10 Healthy Individuals: Implications for Whole Genome–Based Clinical Diagnostics," *Genetics in Medicine* 13, no. 3 (March 2011): 210–17.

16. Tom Chivers, "How the Constant Flow of Data Is Revolutionising Biology," *Independent*, November 13, 2018.

17. Enseqlopedia, "NGS Mapped," http://enseqlopedia.com/ngs-mapped.

18. Eric S. Lander et al., "Initial Sequencing and Analysis of the Human Genome," *Nature* 409, no. 6822 (February 2001): 860–921.

19. Matthew T. Noakes et al., "Increasing the Accuracy of Nanopore DNA Sequencing Using a Time-Varying Cross Membrane Voltage," *Nature Biotechnology* 37, no. 8 (June 2019): 1–6.

20. Tina Woods, "'Longevity' Could Reach Billions in 2019—and Is No Longer Just the Preserve of Billionaires," *Forbes*, January 11, 2019.

21. Prachi Kothiyal et al., "Mendelian Inheritance Errors in Whole Genome Sequenced Trios Are Enriched in Repeats and Cluster Within Copy Number Losses," preprint, submitted December 28, 2017, biorxiv.org/content/10.1101/240424v1.full.

22. Frederick Sanger et al., "Nucleotide Sequence of Bacteriophage φX174 DNA," *Nature* 265, no. 5596 (February 1977): 687–95.

23. H. Richard Johnston et al., "PEMapper and PECaller Provide a Simplified Approach to Whole-Genome Sequencing," *PNAS* 114, no. 10 (2017): e1923–32.

24. Laksshman Sundaram et al., "Predicting the Clinical Impact of Human Mutation with Deep Neural Networks," *Nature Genetics* 50, no. 8 (August 2018): 1161–70.

25. Kary Mullis et al., "Specific Enzymatic Amplification of DNA in Vitro: The Polymerase Chain Reaction," in *Cold Spring Harbor Symposia on Quantitative Biology* (Cold Spring Harbor, NY: Cold Spring Harbor Laboratory, 1986), 51: 263–73.

26. Henrik Stranneheim and Joakim Lundeberg, "Stepping Stones in DNA Sequencing," *Biotechnology* 7, no. 9 (2012): 1063–73.

27. Dmitri Pavlichin and Tsachy Weissman, "The Desperate Quest for Genomic Compression Algorithms," IEEE Spectrum, August 22, 2018, https://spectrum.ieee.org/computing/software/the-desperate-quest-for-genomic-compression-algorithms.

28. Darrel Ince, "The Duke University Scandal—What Can Be Done?" *Significance* 8, no. 3 (2011): 113–15.

29. Kary Mullis, "The Polymerase Chain Reaction," Nobel lecture, December 8, 1993.

30. Kimberly Robasky, Nathan E. Lewis, and George M. Church, "The Role of Replicates for Error Mitigation in Next-Generation Sequencing," *Nature Reviews Genetics* 15 (2014): 56–62.
31. Erwin L. van Dijkl, "The Third Revolution in Sequencing Technology," *Trends in Genetics* 34 , no. 9 (June 2018): 666–81.
32. Bernard Marr, "The Wonderful Ways Artificial Intelligence Is Transforming Genomics and Gene Editing," *Forbes*, November 16, 2018.
33. Frederick Sanger et al., "Use of DNA Polymerase I Primed by a Synthetic Oligonucleotide to Determine a Nucleotide Sequence in Phage fl DNA," *PNAS* 70, no. 4 (April 1973): 1209–13.

Part III

1. Peter M Visscher et al., "10 Years of GWAS Discovery: Biology, Function, and Translation," *American Journal of Human Genetics* 101, no. 1 (July 2017): 5–22.
2. Antonio Regalado, "2017 Was the Year Consumer DNA Testing Blew Up," *MIT Technology Review*, February 12, 2018.
3. Charles Seife, "23andMe Is Terrifying, but Not for the Reasons the FDA Thinks," *Scientific American*, November 27, 2013.
4. Robin P. Smith et al., "Neanderthal Ancestry Inference" (White Paper 23-052015, 23andMe, 2015), https://permalinks.23andme.com/pdf/23 -05_neanderthal_ancestry_inference.pdf.
5. Chun-Xiao Song et al., "5-Hydroxymethylcytosine Signatures in Cell-Free DNA Provide Information About Tumor Types and Stages," *Cell Research* 27, no. 10 (October 2017): 1231–42.
6. Sarah Zhang, "A DNA Company Wants You to Help Catch Criminals," *Atlantic*, March 29, 2019.
7. Leah Eisenstadt, "After a Decade of Genome-Wide Association Studies, a New Phase of Discovery Pushes On," *Broad Institute* (News), August 14, 2017.
8. Carl Zimmer, "Ancestors of Modern Humans Interbred with Extinct Hominins, Study Finds," *New York Times*, March 17, 2016.
9. Lizzie Johnson and Trisha Thadani, "Arrest of Suspected Golden State Killer Through Genealogy Opens 'Pandora's Box,'" *San Francisco Chronicle*, April 29, 2018.
10. Sarah Zhang, "Big Pharma Would Like Your DNA," *Atlantic*, July 28, 2018.
11. Pam Belluck, "Blood Test Might Predict Pregnancy Due Date and Premature Birth," *New York Times*, June 7, 2018.
12. Richard Gray, "Career Opportunities Are at the New Frontier of Genomic Medicine," *BBC Capital*, October 2018.

13. Roy D. Bloom et al., "Cell-Free DNA and Active Rejection in Kidney Allografts," *JASN* 28, no. 7 (2017): 2221–32.
14. Julia Wynn et al., "Clinical Providers' Experiences with Returning Results from Genomic Sequencing: An Interview Study," *BMC Medical Genomics* 11, no. 1 (2018): 45.
15. Gina Kolata, "Clues to Your Health Are Hidden at 6.6 Million Spots in Your DNA," *New York Times*, August 13, 2018.
16. Heather Murphy, "Coming Soon to a Police Station Near You: The DNA 'Magic Box,'" *New York Times*, January 21, 2019.
17. Anne Wojcick, "Consumers Don't Need Experts to Interpret 23andMe Genetic Risk Reports," *STAT*, April 9, 2018.
18. Carl Zimmer, "Deep in Human DNA, a Gift from the Neanderthals," *New York Times*, October 4, 2018.
19. Christopher Mele, "DNA Links Colorado Murders from 34 Years Ago to Inmate," *New York Times*, August 10, 2018.
20. Heather Murphy, "Don't Count on 23andMe to Detect Most Breast Cancer Risks, Study Warns," *New York Times*, April 16, 2019.
21. Astead W. Herndon, "Elizabeth Warren Apologizes to Cherokee Nation for DNA Test," *New York Times*, February 1, 2019.
22. Jonathan Martin, "Elizabeth Warren Releases DNA Results on Native American Ancestry," *New York Times*, October 15, 2018.
23. Pasqualina Colella, Giuseppe Ronzitti, and Federico Mingozzi, "Emerging Issues in AAV-Mediated In Vivo Gene Therapy," *Molecular Therapy: Methods & Clinical Development* 8 (December 2017): 87–104.
24. Jay N. Lozier, "Factor IX Padua: Them That Have, Give," *Blood* 120, no. 23 (November 2012): 4452–53.
25. Matthew Haag, "FamilyTreeDNA Admits to Sharing Genetic Data with F.B.I.," *New York Times*, February 4, 2019.
26. US Food & Drug Association, "FDA Approves Novel Gene Therapy to Treat Patients with a Rare Form of Inherited Vision Loss," FDA News Release, December 19, 2017, www.fda.gov/news-events/press-announcements/fda-approves-novel-gene-therapy-treat-patients-rare-form-inherited-vision-loss.
27. Julie Steenhuysen, "Fetal DNA Tests Prove Highly Accurate but Experts Warn of Exceptions," Reuters, April 1, 2015.
28. James J. Lee et al., "Gene Discovery and Polygenic Prediction from a Genome-Wide Association Study of Educational Attainment in 1.1 Million Individuals," *Nature Genetics* 50, no. 8 (August 2018): 1112–21.
29. Heather Murphy, "Genealogists Turn to Cousins' DNA and Family Trees to Crack Five More Cold Cases," *New York Times*, June 27, 2018.
30. Natalie Ram et al., "Genealogy Databases and the Future of Criminal Investigation," *Science* 360, no. 6393 (June 2018): 1078–79.
31. Heather Murphy, "Genealogy Sites Have Helped Identify Suspects. Now They've Helped Convict One," *New York Times*, July 1, 2019.

32. Amit V. Khera et al., "Genome-Wide Polygenic Scores for Common Diseases Identify Individuals with Risk Equivalent to Monogenic Mutations," *Nature Genetics* 50 (2018): 1219–24.

33. Carrie Louise Hammond, Josh Matthew Willoughby, and Michael James Parker, "Genomics for Paediatricians: Promises and Pitfalls," *Archives of Disease in Childhood* 103, no. 9 (2018): 895–900.

34. Kathryn Gray and Louise E. Wilkins-Haug, "Have We Done Our Last Amniocentesis? Updates on Cell-Free DNA for Down Syndrome Screening," *Pediatric Radiology* 48, no. 4 (2018): 461–70.

35. Lindsey A. George et al., "Hemophilia B Gene Therapy with a High-Specific Activity Factor IX Variant," *New England Journal of Medicine* 377, no. 23 (2017): 2215–27.

36. Michael Makris, "Hemophilia Gene Therapy Is Effective and Safe," *Blood* 131, no. 9 (March 2018): 952–53.

37. Sumathi Reddy, "High Hopes for a Gene Therapy Come with Fears over Cost," *Wall Street Journal*, updated September 24, 2018.

38. Heather Murphy, "How an Unlikely Family History Website Transformed Cold Case Investigations," *New York Times*, October 15, 2018.

39. Simon Erickson, "How Pharmacogenomics Is Improving American Healthcare," *Motley Fool*, August 10, 2018.

40. Thierry VandenDriessche and Marinee K. Chuah, "Hyperactive Factor IX Padua: A Game-Changer for Hemophilia Gene Therapy," *Molecular Therapy* 26, no. 1 (January 2018): 14–16.

41. Erin Aubry Kaplan, "I Don't Need a DNA Test to Tell Me How Black I Am," *New York Times*, April 16, 2019.

42. Nabil Sabri Enattah et al., "Identification of a Variant Associated with Adult-Type Hypolactasia," *Nature Genetics* 30 (February 2002): 233–37.

43. Yaniv Erlich et al., "Identity Inference of Genomic Data Using Long-Range Familial Searches," *Science* 362, no. 6415 (November 2018): 690–94.

44. Amy Harmon, "James Watson Won't Stop Talking About Race," *New York Times*, January 1, 2019.

45. Federico Abascal et al., "Loose Ends: Almost One in Five Human Genes Still Have Unresolved Coding Status," *Nucleic Acids Research* 46, no. 14 (2018): 7070–84.

46. Ramin Nazarian et al., "Melanomas Acquire Resistance to B-RAF(V600E) Inhibition by RTK or N-RAS Upregulation," *Nature* 468, no. 7326 (December 2010): 973–77.

47. Sam Roberts, "Minorities in U.S. Set to Become Majority by 2042," *New York Times*, August 14, 2008.

48. Antonio Regaldo, "More than 26 Million People Have Taken an at-Home Ancestry Test," *MIT Technology Review*, February 11, 2019.

49. Heather Murphy, "Most White Americans' DNA Can Be Identified Through Genealogy Databases," *New York Times*, October 11, 2018.

50. Artur Cideciyan et al., "Mutation-Independent Rhodopsin Gene Therapy by Knockdown and Replacement with a Single AAV Vector," *PNAS* 115, no. 36 (September 2018): e8547–56.

51. Catherine Offord, "MyHeritage Launches Health-Related Genetic Test, Ignites Debate," *Scientist*, July 9, 2019.

52. Carl Zimmer, "Narrower Skulls, Oblong Brains: How Neanderthal DNA Still Shapes Us," *New York Times*, December 13, 2018.

53. Philipp Gunz et al., "Neandertal Introgression Sheds Light on Modern Human Endocranial Globularity," *Current Biology* 29, no. 1 (January 2019): 120–27.

54. Michael D. Gregory et al., "Neanderthal-Derived Genetic Variation Shapes Modern Human Cranium and Brain," *Scientific Reports* 7, no. 6308 (2017).

55. Cheryl Soohoo, "New Blood," *Northwestern Medicine Magazine*, Summer 2018.

56. Thuy T. M. Ngo et al., "Noninvasive Blood Tests for Fetal Development Predict Gestational Age and Preterm Delivery," *Science* 360, no. 6393 (June 2018): 1133–36.

57. H. Christina Fan et al., "Noninvasive Diagnosis of Fetal Aneuploidy by Shotgun Sequencing DNA from Maternal Blood," *PNAS* 105, no. 42 (October 2008): 16266–71.

58. Arlene Weintraub, "Novartis Struggling to Win Payer Coverage for $2.1M Gene Therapy Zolgensma: Analysts," FiercePharma, July 3, 2019.

59. Ann Lin et al., "Off-Target Toxicity Is a Common Mechanism of Action of Cancer Drugs Undergoing Clinical Trials," *Science Translational Medicine* 11, no. 509 (September 2019): eaaw8412.

60. Ann Gibbons, "Oldest *Homo sapiens* Genome Pinpoints Neandertal Input," *Science* 343, no. 6178 (March 2014): 1417.

61. Howard S. Smith, "Opioid Metabolism," *Mayo Clinic Proceedings* 84, no. 7 (July 2009): 613–24.

62. Caroline F. Wright et al., "Paediatric Genomics: Diagnosing Rare Disease in Children," *Nature Reviews Genetics* 19 (May 2018): 253–68.

63. David B Simon et al., "Paracellin-1, a Renal Tight Junction Protein Required for Paracellular Mg2+ Resorption," *Science* 285, no. 5424 (July 1999): 103–6.

64. Federico Rossari, Filippo Minutolo, and Enrico Orciuolo, "Past, Present, and Future of Bcr-Abl Inhibitors: From Chemical Development to Clinical Efficacy," *Journal of Hematology & Oncology* 11 (2018): 84.

65. Laura H. Goetz and Nicholas Schork, "Personalized Medicine: Motivation, Challenges, and Progress," *Fertility and Sterility* 109, no. 6 (June 2018): 952–63.

66. Jonathan P. Jarow, "Personalized Reproductive Medicine: Regulatory Considerations," *Fertility and Sterility* 109, no. 6 (June 2018): 964–67.

67. Christine Birak and Melanie Glanz, "Pharmacies Selling DNA Tests to Help Patients Pick Best Medications," *CBC News*, October 25, 2018.

68. Eugenia Yiannakopoulou, "Pharmacogenomics and Opioid Analgesics: Clinical Implications," *International Journal of Genomics* (2015): 368979, https://doi.org/10.1155/2015/368979.

69. Josh P. Roberts, "Pharmacogenomics: Better Drugs Through Better Screening," *Science* 361, no. 6409 (September 2018): 1396.

70. Jordi Merino and Jose C. Florez, "Precision Medicine in Diabetes: An Opportunity for Clinical Translation," *Annals of the New York Academy of Sciences* 1411, no. 1 (2018): 140–52.

71. Jose C. Florez, "Precision Medicine in Diabetes: Is It Time?," *Diabetes Care* 39, no. 7 (2016): 1085–88.

72. Caroline Ogilvie and Ranjit Akolekar, "Pregnancy Loss Following Amniocentesis or CVS Sampling—Time for a Reassessment of Risk," *Journal of Clinical Medicine* 3, no. 3 (September 2014): 741–46.

73. Jacqueline Mersch et al., "Prevalence of Variant Reclassification Following Hereditary Cancer Genetic Testing," *JAMA* 320, no. 12 (2018): 1266–74.

74. Colleen Flaherty, "Quest for 'Genius Babies'?," *Inside Higher Ed*, May 29, 2013.

75. Susan Gubar, "Raising Awareness of BRCA Mutations," *New York Times*, September 20, 2018.

76. Urko M. Marigorta et al., "Replicability and Prediction: Lessons and Challenges from GWAS," *Trends in Genetics* 34, no. 7 (July 2018): 504–17.

77. Carl Zimmer, "Researchers Explore a Cancer Paradox," *New York Times*, October 22, 2018.

78. National Institutes of Health, "Researchers Find Potential New Gene Therapy for Blinding Disease," NIH News Released, August 20, 2018, www.nih.gov/news-events/news-releases/researchers-find-potential -new-gene-therapy-blinding-disease.

79. Celine Lefebvre, Gabrielle Rieckhof, and Andrea Califano, "Reverse-Engineering Human Regulatory Networks," *WIREs Systems Biology Medicine* 4, no. 4 (2012): 311–25.

80. Alison E. Hall et al., "Risk Stratification, Genomic Data and the Law," *Journal of Community Genetics* 9, no. 3 (2018): 195–99.

81. Richard P. Lifton et al., "Salt and Blood Pressure: New Insight from Human Genetic Studies," in *Cold Spring Harbor Symposia on Quantitative Biology* (2002), 67: 445–50.

82. Gina Kolata, "Scientists Designed a Drug for Just One Patient. Her Name Is Mila," *New York Times*, October 9, 2019.

83. Heather Murphy, "She Helped Crack the Golden State Killer Case. Here's What She's Going to Do Next," *New York Times*, August 29, 2018.

84. Corey Kilgannon, "She Was Left in a Bag as a Newborn. DNA Testing Helped Her Understand Why," *New York Times*, May 22, 2019.

85. Iñigo Martincorena et al., "Somatic Mutant Clones Colonize the Human Esophagus with Age," *Science* 23, no. 6417 (October 2018): 911–17.

86. Heather Murphy, "Sooner or Later Your Cousin's DNA Is Going to Solve a Murder," *New York Times*, April 25, 2019.

87. Aldamaria Puliti et al., "Teaching Molecular Genetics: Chapter 4—Positional Cloning of Genetic Disorders," *Pediatric Nephrology* 22, no. 12 (2007): 2023–29.

88. Matthew Warren, "The Approach to Predictive Medicine That Is Taking Genomics Research by Storm," *Nature* 562 (2018): 181–83.

89. Sheryl Stolberg, "The Biotech Death of Jesse Gelsinger," *New York Times*, November 28, 1999.

90. Kelly Grant, "The Birth of a Revolution in Prenatal Screening," *Globe and Mail*, January 11, 2018.

91. Antonio Regalado and Brian Alexander, "The Citizen Scientist Who Finds Killers from Her Couch," *MIT Technology Review*, June 22, 2018.

92. Kay Prüfer et al., "The Complete Genome Sequence of a Neanderthal from the Altai Mountains," *Nature* 505, no. 7481 (January 2014): 43–9.

93. Lap-Chee Tsui and Ruslan Dorfman, "The Cystic Fibrosis Gene: A Molecular Genetic Perspective," *Cold Spring Harbor Perspectives in Medicine* 3, no. 2 (February 2013): a009472.

94. Sriram Sankararaman et al., "The Genomic Landscape of Neanderthal Ancestry in Present-Day Humans," *Nature* 507, no. 7492 (March 2014): 354–57.

95. Joshua L. Krieger et al., "The Impact of Personal Genomics on Risk Perceptions and Medical Decision-Making," *Nature Biotechnology* 34 (September 2016): 912–18.

96. Stephen S. Rich and William T. Cefalu, "The Impact of Precision Medicine in Diabetes: A Multidimensional Perspective," *Diabetes Care* 39, no. 11 (2016): 1854–57.

97. Brian Resnick, "The Limits of Ancestry DNA Tests, Explained," *Vox* .com, January 28, 2019.

98. Francis S. Collins and Scott Gottlieb, "The Next Phase of Human Gene-Therapy Oversight," *New England Journal of Medicine* 379, no. 15 (October 2018): 1393–95.

99. Michael D. Gallagher and Alice S. Chen-Plotkin, "The Post-GWAS Era: From Association to Function," *American Journal of Human Genetics* 102, no. 5 (May 2018): 717–30.

100. Andrea Califano and Mariano J. Alvarez, "The Recurrent Architecture of Tumour Initiation, Progression and Drug Sensitivity," *Nature Reviews Cancer* 17, no. 2 (February 2017): 116–30.

101. Gina Kolata, "The Results of Your Genetic Test Are Reassuring. But That Can Change," *New York Times*, October 16, 2018.

102. Erika Check Hayden, "The Rise and Fall and Rise Again of 23andMe," *Nature*, October 11, 2017.

103. Tom Higham et al., "The Timing and Spatiotemporal Patterning of Neanderthal Disappearance," *Nature* 512 (2014): 306–9.

104. Gina Kolata, "They Thought Hemophilia Was a 'Lifelong Thing,' They May Be Wrong," *New York Times*, August 13, 2018.

105. Daniel M. Jordan and Ron Do, "Using Full Genomic Information to Predict Disease: Breaking Down the Barriers Between Complex and Mendelian Diseases," *Annual Review of Genomics and Human Genetics* 19 (2018): 289–301.

106. Susana Salceda et al., "Validation of a Rapid DNA Process with the RapidHIT® ID System Using GlobalFiler® Express chemistry, a Platform Optimized for Decentralized Testing Environments," *Forensic Science International: Genetics* 28 (May 2017): 21–34.

107. Jun Gong, "Value-Based Genomics," *Oncotarget* 9, no. 21 (2018): 15792–815.

108. Julie O. Culver et al., "Variants of Uncertain Significance in BRCA Testing: Evaluation of Surgical Decisions, Risk Perception, and Cancer Distress," *Clinical Genetics* 84, no. 5 (November 2013): 464–72.

109. Michael N. Weedon, "Very Rare Pathogenic Genetic Variants Detected by SNP-Chips Are Usually False Positives: Implications for Direct-to-Consumer Genetic Testing," preprint, submitted July 9, 2019, www.biorxiv.org/content/10.1101/696799v1.

110. Elizabeth Joh, "Want to See My Genes? Get a Warrant," *New York Times*, June 11, 2019.

111. Claudia Dreifus, "What Did Neanderthals Leave to Modern Humans? Some Surprises," *New York Times*, January 20, 2017.

112. Oscar Schwatz, "What Does It Mean to Be Genetically Jewish?" *Guardian*, June 13, 2019.

113. Sarah Zhang, "What Spotify's DNA-Test Playlist Gets Wrong About Genetic Ancestry," *Atlantic*, September 25, 2018.

114. Sarah Zhang, "When a DNA Test Shatters Your Identity," *Atlantic*, July 17, 2018.

115. Shaun Raviv and Mosaic, "When Genetic Diseases Threaten Patient Privacy," *Atlantic*, July 14, 2018.

116. Peter M. Gayed, "Why Some People Prefer Pickle Juice: The Research of Dr. Richard P. Lifton," *Yale Journal of Biology and Medicine* 80 (2007): 159–63.

117. Sabrina Tavernise, "Why the Announcement of a Looming White Minority Makes Demographers Nervous," *New York Times*, November 22, 2018.

118. Brett Molina, "Woman Discovers She Has 29 Siblings After Taking DNA Test. And Counting," *USA Today*, April 8, 2019.

119. Robin McKie, "Woman Who Inherited Fatal Illness to Sue Doctors in Groundbreaking Case," *Guardian*, November 25, 2018.

120. Paolo Simioni et al., "X-Linked Thrombophilia with a Mutant Factor IX (Factor IX Padua)," *New England Journal of Medicine* 361 (2009): 1671–75.
121. Carl Zimmer, "Years of Education Influenced by Genetic Makeup, Enormous Study Finds," *New York Times*, July 23, 2018.

Part IV

1. Diana Kwon, "A Brief Guide to the Current CRISPR Landscape," *Scientist*, July 15, 2019.
2. Natalie D. Halbert and James N. Derr, "A Comprehensive Evaluation of Cattle Introgression into US Federal Bison Herds," *Journal of Heredity* 98, no. 1 (2007): 1–12.
3. "A CRISPR Definition of Genetic Modification" (editorial), *Nature Plants* 4 (2018): 233.
4. Kyos Kyrou et al., "A CRISPR-Cas9 Gene Drive Targeting Doublesex Causes Complete Population Suppression in Caged *Anopheles gambiae* Mosquitoes," *Nature Biotechnology* 36, no. 11 (December 2018): 1062–66.
5. Martin Jinek et al., "A Programmable Dual-RNA–Guided DNA Endonuclease in Adaptive Bacterial Immunity," *Science* 337, no. 6096 (August 2012): 816–21.
6. Sharon Begley, "All You Need to Know for Round 2 of the CRISPR Patent Fight," STAT, April 30, 2018.
7. Linda Goodman and Elinor K. Karlsson, "America's Lost Dogs," *Science* 361, 6397 (July 2018): 27–28.
8. Jane J. Lee, "American Dog Breeds Hail From Pre-Columbian Times," *National Geographic*, July 9, 2013.
9. Jane E. Brody, "Are G.M.O. Foods Safe?," *New York Times*, April 23, 2018.
10. Michael Gerson, "Are You Anti-GMO? Then You're Anti-Science, Too," *Washington Post*, May 3, 2018.
11. Julie M. Crudele and Jeffrey S. Chamberlain, "Cas9 Immunity Creates Challenges for CRISPR Gene Editing Therapies," *Nature Communications* 9, no. 3497 (2018).
12. Casey Smith, "Cats Domesticated Themselves, Ancient DNA Shows," *National Geographic*, June 19, 2017.
13. Ana Falcon et al., "CCR5 Deficiency Predisposes to Fatal Outcome in Influenza Virus Infection," *Journal of General Virology* 96, no. 8 (2015): 2074–78.
14. Marilynn Marchione, "Chinese Researcher Claims First Gene-Edited Babies," *Washington Post*, November 26, 2018.
15. Sarah Zhang, "Chinese Scientists Are Outraged by Reports of Gene-Edited Babies," *Atlantic*, November 27, 2018.
16. Samuel K. Wasser et al., "Combating Transnational Organized Crime by Linking Multiple Large Ivory Seizures to the Same Dealer," *Science Advances* 4, no. 9 (2018): eaat0625.

17. Mark Lynas, "Confessions of an Anti-GMO Activist," *Wall Street Journal*, June 22, 2018.

18. Heidi Ledford, "CRISPR Babies: When Will the World Be Ready?," *Nature* 570, no. 7761 (2019): 293–96.

19. Sharon Begley, "CRISPR Cures Inherited Disorder in Mice," STAT, October 9, 2018.

20. Luciano A. Marraffini and Erik J Sontheimer, "CRISPR Interference Limits Horizontal Gene Transfer in Staphylococci by Targeting DNA," *Science* 322, no. 5909 (December 2008): 1843–45.

21. Gavin J. Knott and Jennifer A. Doudna, "CRISPR-Cas Guides the Future of Genetic Engineering," *Science* 361, no. 6405 (2018): 866–69.

22. Philip F. Thomsen et al., "Detection of a Diverse Marine Fish Fauna Using Environmental DNA from Seawater Samples," *PLoS ONE* 7, no. 8 (August 2012).

23. Sarah Zhang, "Does GMO Labeling Actually Increase Support for GMOs?," *Atlantic*, June 27, 2018.

24. Edward Lanphier et al., "Don't Edit the Human Germline," *Nature* 519, no. 7544 (March 2015): 410–11.

25. Harris A. Lewin et al., "Earth BioGenome Project: Sequencing Life for the Future of Life," *PNAS* 115, no. 17 (April 2018): 4325–33.

26. Eric J. B. von Wettberg et al., "Ecology and Genomics of an Important Crop Wild Relative as a Prelude to Agricultural Innovation," *Nature Communications* 9 (2018): 649.

27. Hiroshi Nishimasu et al., "Engineered CRISPR-Cas9 Nuclease with Expanded Targeting Space," *Science* 361, no. 6408 (2018): 1259–62.

28. James A. Cotton, "Eradication Genomics—Lessons for Parasite Control," *Science* 361, no. 6398 (July 2019): 130–31.

29. Juliane Kaminski et al., "Evolution of Facial Muscle Anatomy in Dogs," *PNAS* 116, no. 29 (July 2019): 14677–81.

30. Broad Communications, "For Journalists: Statement and Background on the Crispr Patent Decision," updated September 10, 2018, by Lee McGuire (CCO), Broad Institute of MIT and Harvard, www.broadinstitute.org /crispr/journalists-statement-and-background-crispr-patent-process.

31. Ute V. Solloch et al., "Frequencies of Gene Variant *CCR5* Δ32 in 87 Countries Based on Next Generation Sequencing of 1.3 Million Individuals Sampled from 3 National DKMS Donor Centers," *Human Immunology* 78, nos. 11–12 (2017): 710–17.

32. Amy Harmon, "G.M.O. Foods Will Soon Require Labels. What Will the Labels Say?," *New York Times*, May 12, 2018.

33. Pam Belluck, "Gene-Edited Babies: What a Chinese Scientist Told an American Mentor," *New York Times*, April 14, 2019.

34. Norman F. Weeden, "Genetic Changes Accompanying the Domestication of *Pisum sativum*: Is There a Common Genetic Basis to the

'Domestication Syndrome' for Legumes?," *Annals of Botany* 100, no. 5 (October 2007): 1017–25.

35. Steve Connor, "Genetically Modified Salmon Becomes First to Be Approved for Human Consumption—but It Won't Have to Be Labeled as GM," *Independent*, November 19, 2015.

36. Fred W. Allendorf, Paul A. Hohenlohe, and Gordon Luikart, "Genomics and the Future of Conservation Genetics," *Nature Reviews Genetics* 11 (2010): 697–709.

37. José R. Dinneny, "Getting It Right on GMOs," *Science* 360, no. 6396 (June 2018): 1407.

38. Peter Beetham, "GMOs Are Not Agriculture's Future—Biotech Is," *Scientific American* (blog), September 5, 2018.

39. David Ropeik, "Golden Rice Opponents Should Be Held Accountable for Health Problems Linked to Vitamin A Deficiency," *Scientific American* (blog), March 15, 2014.

40. Susie Nelson, "How Almonds Went from Deadly to Delicious," *NPR*, June 13, 2019.

41. Jon Cohen, "How the Battle Lines over CRISPR Were Drawn," *Science*, February 15, 2017.

42. Pam Belluck, "How to Stop Rogue Gene-Editing of Human Embryos?," *New York Times*, January 24, 2019.

43. Avery C. Rossidis et al., "In Utero CRISPR-Mediated Therapeutic Editing of Metabolic Genes," *Nature Medicine* 24 (October 2018): 1513–18.

44. Deepthi Alapati et al., "In Utero Gene Editing for Monogenic Lung Disease," *Science Translational Medicine* 11, 488 (2019): eaav8375.

45. David Ewing Duncan, "Inside the Very Big, Very Controversial Business of Dog Cloning," *Vanity Fair*, August 7, 2018.

46. Ricki Lewis, "Is CRISPR Gene Editing Doomed, Even as Gene Therapy Enters the Clinic?," *PLOS DNA Sciences Blog*, August 9, 2018.

47. Le Cong et al., "Multiplex Genome Engineering Using CRISPR/Cas Systems," *Science* 339, no. 6121 (February 2013): 819–23.

48. University of Pennsylvania, "NY-ESO-1-redirected CRISPR (TCRendo and PD1) Edited T Cells (NYCE T Cells)" clinical trial, last updated March 4, 2019, ClinicalTrials.gov Identifier NCT03399448.

49. Lia Moses, Steve Niemi, and Elinor Karlsson, "Pet Genomics Medicine Runs Wild," *Nature*, July 25, 2018.

50. Vijaya L. Simhadri et al., "Prevalence of Pre-existing Antibodies to CRISPR-Associated Nuclease Cas9 in the USA Population," *Molecular Therapy—Methods & Clinical Development* 10 (June 2018): 105–12.

51. Anna V. Kukekova et al., "Red Fox Genome Assembly Identifies Genomic Regions Associated with Tame and Aggressive Behaviours," *Nature Ecology and Evolution* 2 (August 2018): 1479–91.

52. Catherine Offord, "Researchers Launch First Study of In Vivo CRISPR Therapy in Humans," *Scientist*, July 26, 2019.

53. Jef Akst, "Researchers Track Sharks and Whales Using DNA in Seawater Samples," *Scientist*, January 1, 2019.

54. Prashant Mali et al., "RNA-Guided Human Genome Engineering via Cas9," *Science* 339, no. 6121 (February 2013): 823–26.

55. Rob Stein, "Scientists Release Controversial Genetically Modified Mosquitoes in High-Security Lab," *NPR* (blog), February 20, 2019.

56. Gina Kolata, "South Korean Scientists Clone Man's Best Friend, a First," *New York Times*, August 3, 2005.

57. Pam Belluck, "Stanford Clears Professor of Helping with Gene-Edited Babies Experiment," *New York Times*, April 16, 2019.

58. US Food & Drug Administration, "Statement from FDA Commissioner Scott Gottlieb, M.D., on Continued Efforts to Advance Safe Biotechnology Innovations, and the Deactivation of an Import Alert on Genetically Engineered Salmon," FDA Statement, March 8, 2019.

59. "Super-Tomato Shows What Plant Scientists Can Do" (editorial), *Nature* 562, no. 7725 (2018): 8.

60. Veronique Greenwood, "Taming the Groundcherry: With Crispr, a Fussy Fruit Inches Toward the Supermarket," *New York Times*, October 8, 2018.

61. Máire Ní Leathlobhair et al., "The Evolutionary History of Dogs in the Americas," *Science* 361, no. 6397 (July 2018): 81–85.

62. Katie O'Reilly, "The Great GMO Switcheroo," *National Magazine of the Sierra Club*, July 3, 2018.

63. James Gorman, "The Lost Dogs of America," *New York Times*, July 5, 2018.

64. Claudio Ottoni et al., "The Palaeogenetics of Cat Dispersal in the Ancient World," *Nature Ecology & Evolution* 1, no. 0139 (2017).

65. Anna Janská, "The Role of the Testa During Development and in Establishment of Dormancy of the Legume Seed," *Frontiers in Plant Science* 5, no. 351 (July 2014): 815–29.

66. Jenny Spli, "Time to Get Rid of 'Non-GMO' Argues 'Biotech-Backed' Group," *Forbes*, September 30, 2018.

67. Carl Zimmer, "What Is a Genetically Modified Crop? A European Ruling Sows Confusion," *New York Times*, July 27, 2018.

68. Katherine J. Wu, "What the Fox Genome Tells Us About Domestication," *Smithsonian*, August 14, 2018.

69. Sarah Zhang, "What Vets Think of '23andMe for Dogs,'" *Atlantic*, November 12, 2018.

70. James Gorman, "Why Are These Foxes Tame? Maybe They Weren't So Wild to Begin With," *New York Times*, December 3, 2019.

Part V

1. Egan Molteny, "A Controversial Fertility Treatment Gets Its First Test," *WIRED*, January 28, 2019.
2. Kristien Hens, Wybo Dondorp, and Guido de Wert, "A Leap of Faith? An Interview Study with Professionals on the Use of Mitochondrial Replacement to Avoid Transfer of Mitochondrial Diseases," *Human Reproduction* 30, no. 5 (March 2015): 1256–62.
3. Kevin N. Laland, "A Pair of Evolutionary Biologists Takes a Closer Look at Nongenetic Inheritance," *Science*, June 26, 2018.
4. Alice Park, "An Experimental Procedure Could Help More Families Have Healthy Babies. But It's Not Allowed in the US," *Time*, January 3, 2019.
5. Jacques Cohen et al., "Birth of Infant After Transfer of Anucleate Donor Oocyte Cytoplasm into Recipient Eggs," *Lancet* 350, no. 9072 (July 1997): 186–87.
6. Alison Abbott, "Can Epigenetics Help Verify the Age Claims of Refugees?," *Nature* 561 (2018): 15.
7. Adam Field et al., "DNA Methylation Clocks in Aging: Categories, Causes, and Consequences," *Molecular Cell* 71, no. 6 (September 2018): 882–95.
8. Chongyuan Luo, Petra Hajkova, and Joseph R. Ecker, "Dynamic DNA Methylation: In the Right Place at the Right Time," *Science* 361, no. 6409 (2018): 1336–40.
9. Masoumeh Fardi, Saeed Solali, and Majid Farshdousti Hagh, "Epigenetic Mechanisms as a New Approach in Cancer Treatment: An Updated Review," *Genes & Diseases* 5, no. 4 (December 2018): 304–11.
10. Amit Berson et al., "Epigenetic Regulation in Neurodegenerative Diseases," *Trends in Neuroscience* 41, no. 9 (September 2018): 587–98.
11. Martina Collotta, Pier A. Bertazzi, and Valentina Bollati, "Epigenetics and Pesticides," *Toxicology* 307 (2013): 35–41.
12. Deqing Hu and Ali Shilatifard, "Epigenetics of Hematopoiesis and Hematological Malignancies," *Genes and Development* 30 (2016): 2021–41.
13. Steven Henikoff and John M. Greally, "Epigenetics, Cellular Memory and Gene Regulation," *Current Biology* 26 (2016): R641–66.
14. Jose R. Blesa, Julio Tudela, and Justo Aznar, "Ethical Aspects of Nuclear and Mitochondrial DNA Transfer," *Linacre Quarterly* 83, no. 2 (2016): 179–91.
15. James D. Watson and Francis H. C. Crick, "Genetical Implications of the Structure of Deoxyribonucleic Acid," *Nature* 171, no. 4361 (March 1953): 964–67.
16. Carl Zimmer, "Genetically Modified People Are Walking Among Us," *New York Times*, December 1, 2018.
17. Denise P. Barlow and Marisa S. Bartolomei, "Genomic Imprinting in Mammals," *Cold Spring Harbor Perspectives in Biology* 6 (2014): a018382.

18. Olga Khazan, "Inherited Trauma Shapes Your Health," *Atlantic*, October 18, 2018.
19. International Human Genome Sequencing Consortium, "Initial Sequencing and Analysis of the Human Genome," *Nature* 409, no. 6822 (February 2001): 860–921.
20. Shuvra Shekhar Roy, Ananda Kishore Mukherjee, and Shantanu Chowdhury, "Insights About Genome Function from Spatial Organization of the Genome," *Human Genetics* 12, no. 1 (2018): 8.
21. Bluma J. Lesch et al., "Intergenerational Epigenetic Inheritance of Cancer Susceptibility in Mammals," eLIFE 8 (2019): e39380.
22. Anastasia I Ryzhkova et al., "Mitochondrial Diseases Caused by mtDNA Mutations: A Mini-Review," *Therapeutics and Clinical Risk Management* 14 (2018): 1933–42.
23. Robert W. Taylor and Doug M. Turnbull, "Mitochondrial DNA Mutations in Human Disease," *Nature Review Genetics* 6, no. 5 (2005): 389–402.
24. Daniel F. Bogenhagen, "Mitochondrial DNA Nucleoid Structure," *Biochimica et Biophysica Acta* 1819, nos. 9–10 (September–October 2012): 914–20.
25. Jonathan R. Friedman and Jodi Nunnari, "Mitochondrial Form and Function," *Nature* 505, no. 7483 (January 2014): 335–43.
26. James D. Watson and Francis H. C. Crick, "Molecular Structure of Nucleic Acids: A Structure for Doxyribose Nucleic Acid," *Nature* 171, no. 4356 (April 1953): 737–38.
27. Jan A. M. Smeitink et al., "Nuclear Genes of Human Complex I of the Mitochondrial Electron Transport Chain: State of the Art," *Human Molecular Genetics* 7, no. 10 (September 1998): 1573–79.
28. John D. Loike, "Opinion: The New Frontiers of Epigenetics," *Scientist*, November 12, 2018.
29. Emily Mullin, "Pregnancy Reported in the First Known Trial of 'Three-Person IVF' for Infertility," STAT, January 24, 2019.
30. Yuval Dor and Howard Cedar, "Principles of DNA Methylation and Their Implications for Biology and Medicine," *Lancet* 392, no. 10149 (September 2018): P777–86.
31. Lyndsey Craven et al., "Pronuclear Transfer in Human Embryos to Prevent Transmission of Mitochondrial DNA Disease," *Nature* 465, no. 7294 (May 2010): 82–85.
32. Gavin Kelsey, Oliver Stegle, and Wolf Reik, "Single-Cell Epigenomics: Recording the Past and Predicting the Future," *Science* 358, no. 6359 (2017): 69–75.
33. María Isabel Vaquero-Sedas and Miguel Ángel Vega-Palas, "Targeting Cancer Through the Epigenetic Features of Telomeric Regions," *Trends in Cell Biology* 29, no. 4 (January 2019): 281–90.

34. Jee-Yeon Hwang, Kelly A. Aromolaran, and R. Suzanne Zukin, "The Emerging Field of Epigenetics in Neurodegeneration and Neuroprotection," *Nature Reviews Neuroscience* 18 (June 2017): 347–61.
35. Michael A. Reid, Ziwei Dai, and Jason W. Locasale, "The Impact of Cellular Metabolism on Chromatin Dynamics and Epigenetics," *Nature Reviews Cell Biology* 19, no. 11 (2017): 1298–306.
36. Andrew P. Feinberg, "The Key Role of Epigenetics in Human Disease Prevention and Mitigation," *New England Journal of Medicine* 378 (April 2018): 14.
37. Frederick Sanger et al., "Use of DNA Polymerase I Primed by a Synthetic Oligonucleotide to Determine a Nucleotide Sequence in Phage fl DNA," *PNAS* 70, no. 4 (April 1973): 1209–13.
38. Tahsin Stefan Barakat and Joost Gribnau, "X Chromosome Inactivation and Embryonic Stem Cells," in *Madame Curie Bioscience Database* [Internet] (Austin, TX: Landes Bioscience, 2000–2013).
39. Rafael Galupa and Edith Heard, "X-Chromosome Inactivation: A Crossroads Between Chromosome Architecture and Gene Regulation," *Annual Review of Genetics* 52 (2018): 535–66.

ABOUT THE AUTHOR

Dr. Joshua Z. Rappoport received a bachelor's degree in biology from Brown University and then went on to earn a PhD from the Program in Mechanisms of Disease and Therapeutics at the Mount Sinai School of Medicine Graduate School of Biological Sciences of New York University. Dr. Rappoport then went on to perform postdoctoral work at The Rockefeller University in New York City in the Laboratory of Cellular Biophysics. He was later recruited as a faculty member in the School of Biosciences at the University of Birmingham in England.

In 2014 Dr. Rappoport returned to the United States to serve as the director of the Center for Advanced Microscopy and Nikon Imaging Center at Northwestern University Feinberg School of Medicine and as a faculty member in the Department of Molecular and Cell Biology. After five years at Northwestern University, Dr. Rappoport moved to Boston College, where, as of March 1, 2019, he is the executive director of research infrastructure.